POPULATION DYNAMICS

Alternative Models

PHYSIOLOGICAL ECOLOGY

A Series of Monographs, Texts, and Treatises

EDITED BY

T. T. KOZLOWSKI

University of Wisconsin
Madison, Wisconsin

T. T. Kozlowski. Growth and Development of Trees, Volumes I and II – 1971

Daniel Hillel. Soil and Water: Physical Principles and Processes, 1971

J. Levitt. Responses of Plants to Environmental Stresses, 1972

V. B. Youngner and C. M. McKell (Eds.). The Biology and Utilization of Grasses, 1972

T. T. Kozlowski (Ed.). Seed Biology, Volumes I, II, and III – 1972

Yoav Waisel. Biology of Halophytes, 1972

G. C. Marks and T. T. Kozlowski (Eds.). Ectomycorrhizae: Their Ecology and Physiology, 1973

T. T. Kozlowski (Ed.). Shedding of Plant Parts, 1973

Elroy L. Rice. Allelopathy, 1974

T. T. Kozlowski and C. E. Ahlgren (Eds.). Fire and Ecosystems, 1974

J. Brian Mudd and T. T. Kozlowski (Eds.). Responses of Plants to Air Pollution, 1975

Rexford Daubenmire. Plant Geography, 1978

John G. Scandalios (Ed.), Physiological Genetics, 1979

Bertram G. Murray, Jr. Population Dynamics: Alternative Models, 1979

In Preparation

J. Levitt. Responses of Plants to Environmental Stress, 2nd Edition. Volume I: Chilling, Freezing, and High Temperature Stresses, 1979, Volume II: Water, Radiation, Salt, and Miscellaneous Stresses, 1980

James A. Larsen. The Boreal Ecosystem, 1980

POPULATION DYNAMICS

ALTERNATIVE MODELS

Bertram G. Murray, Jr.

University College
Rutgers University
New Brunswick, New Jersey

ACADEMIC PRESS

A Subsidiary of Harcourt Brace Jovanovich, Publishers

New York   London   Toronto   Sydney   San Francisco   1979

ACADEMIC PRESS, INC.
111 Fifth Avenue, New York, New York 10003

United Kingdom Edition published by
ACADEMIC PRESS, INC. (LONDON) LTD.
24/28 Oval Road, London NW1 7DX

Library of Congress Cataloging in Publication Data

Murray, Bertram G
 Population dynamics.

 (Physiological ecology series)
 Bibliography: p.
 Includes index.
 1. Population biology. 2. Population biology——
Mathematical models. I. Title.
QH352.M97 574.5'24 79–52790
ISBN 0–12–511750–4

PRINTED IN THE UNITED STATES OF AMERICA

79 80 81 82   9 8 7 6 5 4 3 2 1

CONTENTS

Preface   **vii**

1 Introduction   **1**

    The Population Problem   3
    Definitions   11
    Questions   15

2 Population Mathematics   **17**

    Exponential and Logistic Equations   19
    Lotka's Equations   20
    The Projection Matrix   34
    The Effect of Age Distribution   35

3 Limitations to Population Growth   **37**

    Space as a Limiting Factor   38
    Food and Predation as Limiting Factors   45
    Time as a Limiting Factor   59
    The Role of "Density-Dependent" Factors   66

Distribution and Abundance   70
Evaluation of Older Theories   73

4 Population Dynamics and Natural Selection   76

A Theory of Evolution of Life History Patterns   82
Previous Theories   107
Clutch Size and Population Dynamics   121
r-Selection and K-Selection   122

5 Predation and Maximum Sustainable Yield   133

Effects of Predation on Prey Populations   135
Complicating Factors   147
Laboratory Experiments   151
Prudent Predation and Maximum Sustainable Yield   160
Control of Pest Populations   161

6 Competition   164

Competition without Shortages   181

7 The Dynamics of Populations   183

Bibliography   189

Subject Index   205

PREFACE

The examination, and, where need be, revision of our fundamental premises is a task of a wholly different order from that of rearing upon these premises a structure of logical argumentation. . . . Most of us are held back by our preconceived, intuitive judgments, which, blindly entertained, blind us also against the recognition of possible alternatives.

Lotka, 1925

The central topic of ecological interest in the 1950s was population dynamics. The peak occurred in 1954 with the publication of books by H. G. Andrewartha and L. C. Birch, by David Lack, and by Umberto D'Ancona; lengthy papers by A. J. Nicholson, by L. C. Cole, and by W. E. Ricker; a symposium on population cycles; and many additional papers, representing a diversity of views. In 1957 the Cold Spring Harbor Symposium of Quantitative Biology was devoted to population ecology and demography. At this meeting the proponents of opposing theories faced each other but failed to resolve their differences. The majority view held that population growth rates and population sizes were in one way or another regulated by density-dependent factors. The minority view contended that populations could not grow beyond limits set by the availability of resources or of time in which to grow, both of which were often influenced by the weather. Without resolving the differences that gave rise to these opposing ideas, ecologists shifted their attention to community ecology—structure, stability, diversity, competition, and coexistence. The debate resurfaced briefly in 1970 when an international symposium convened in the Netherlands to discuss the "Dynamics of Numbers in Populations," but the majority view held firm.

Unfortunately, ecology textbooks only skim the issues of the 1950s' discussion of population dynamics. The usual treatment is to present the exponential and logistic equations, to mention density-dependent and density-independent factors, and to present examples that are said to support the notion that the growth in numbers of a population is regulated by density-dependent factors. But the significance of particular data cannot correctly be evaluated unless the predictions of alternative theories are explicitly clear. The minority view of population dynamics, however, is not clearly stated by its critics or by textbook authors, and, thus, the alternative models proposed to explain the dynamics of populations have not been justly evaluated. Perhaps the minority view has been rejected prematurely.

In this book I return to the central ecological problem of the 1950s. My purpose is to establish a theoretical framework for thinking about population dynamics different from the "density-dependent regulation" paradigm, which prevails at this time. It is not my purpose to review the many interpretations of population dynamics that have already been reviewed so many times. I do not attempt to sort out the arguments for and against previous views, and I do not include many papers written within the "density-dependent regulation" paradigm whose purpose was to fill in the details of that paradigm. I do attempt to present an explicit view of population dynamics, always keeping in mind the controversies surrounding past discussions.

I believe that past controversies were more the result of ambiguous presentation than of legitimate differences of interpretation. Different terms may be used interchangeably by one author and as contrasting alternatives by another. Often, assumptions are unstated, or perhaps unrealized, leaving the reader to puzzle out for himself the logical argument leading to an author's conclusions. Therefore, I have attempted to avoid these problems in communication by defining terms as I intend to use them, by identifying sources of ambiguity, by stating my assumptions, and by framing the questions I propose to answer. This is done throughout the book, but Chapters 1 and 2 provide a general introduction to the terminology, the mathematical background, and the philosophical approach that lie behind the theoretical development. There follows in Chapter 3 a series of models accounting for variations in population growth rates, sizes, and fluctuations, and in Chapter 4, a model accounting for the evolution of life history patterns. A more detailed examination of the effects of predation on prey populations, especially with respect to determining a prey population's maximum sustainable yield, is taken up in Chapter 5. In Chapter 6, interspecific competition theory is put in terms of the population dynamics models presented in Chapter 3. Chapter 7 presents a summary discussion of population dynamics as it has been developed in this book.

I am most grateful to H. G. Andrewartha and L. C. Birch for writing their book, "The Distribution and Abundance of Animals," which has provided a continuing stimulus to think about the dynamics of populations during the past two decades when it was my misfortune to fail to find a colleague who could discuss population dynamics outside the "density-dependent regulation" paradigm. As an ornithologist, I am grateful to the late David Lack for his series of books, which synthesized the literature on avian population biology and always focused on the important problems of population dynamics. Several friends read and commented on one or more versions of the manuscript, reducing, if not eliminating, the number of errors: Lawrence J. Corwin (Chapter 2), Kenneth A. Crossner (an early version of the manuscript), Erica Dunn (Chapter 4), Michael Gochfeld (an early version), David J. T. Hussell (Chapter 4), Charles Leck (an early version), Steward T. A. Pickett (the final version), and William Shields (earlier versions). Comments by Robert M. Mengel and Marion Anne Jenkinson, made in response to an oral presentation of mine on the evolution of clutch size, led to substantive additions to the text. Finally, my wife, Patti, provided moral support and patiently put up with me in my struggle to put ideas on paper.

Bertram G. Murray, Jr.

1 | INTRODUCTION

An understanding of the dynamics of populations has been attempted by naturalists (e.g., Lack, 1954, 1966), by experimentalists (e.g., Nicholson, 1933, 1954a), and by mathematicians (e.g., Lotka, 1925). This three-pronged approach has produced the prevailing view that the size of populations is usually regulated by density-dependent factors. The size of many populations fluctuates between limits around some mean value, frequently characterized as the carrying capacity of the habitat. As the population grows, density-dependent factors, such as intraspecific competition for resources, predation, or disease, increase in intensity and lower the population's growth rate, bringing the population into balance with its environment. In some populations, density-dependent factors are inadequate, nonfunctional, or nonexistent; thus, periods of exponential growth are followed by population crashes.

To generalize the prevailing view in this way is hazardous, as a careful reading of primary sources shows. The contributions of Nicholson (1933, 1954a,b, 1958a,b), Nicholson and Bailey (1935), Solomon (1949, 1957, 1964, 1969), Lack (1954, 1966), Ricker (1954), Huffaker (1958a), Chitty (1960, 1967), Wynne-Edwards (1962), and Huffaker and Messenger (1964a), among others, provide a diversity of interpretations of population dynamics, particularly with respect to the nature and action of the regulating factors. Nevertheless, these authors agree that density-dependent factors play an important role in regulating a population's size.

What is clearly not the prevailing view is the notion that the size of a population is not regulated but is limited by the availability of resources or by the length of the reproductive period, these often being affected by factors occurring independently of density, such as the weather (Thompson, 1939, 1956; Schwerdtfeger, 1958). This was most eloquently and thoroughly discussed by Andrewartha and Birch (1954).

Milne (1957, 1958a, 1962) proposed a theory that combines aspects of the opposing sides, suggesting that intraspecific competition (a "perfect" density-dependent factor) sets the upper limit but that "imperfect" density-dependent factors and density-independent factors usually combine to limit population size before intraspecific competition becomes effective.

The distinction between these theories is not always kept clear. Indeed, the terms "regulation" and "limitation" are frequently used interchangeably. In the first, however, density-dependent factors are said to determine numbers by regulating the growth rate of a population such that its size varies slightly about some mean value (the habitat's carrying capacity) determined by the quality of the habitat. The Andrewartha and Birch theory of population limitation does not state that "density-independent factors regulate population size," as suggested by some of its critics (e.g., Kuenen, 1958; Klomp, 1962), but that (1) continuous population growth is limited by the availability of resources and of time in which to grow, and (2) the existing density-dependent factors affecting birth and death rates are not effective enough to regulate the population's size.

The basis of much mathematical population theory is the logistic equation of Verhulst (1838), which was independently derived by Pearl and Reed (1920). From the logistic equation has developed the well-known competition equations of Volterra (1931) and Lotka (1932) and the notion of different selection pressures on growth rates, called r-selection and K-selection (MacArthur and Wilson, 1967). The logistic equation survives as the most generally accepted mathematical representation of population growth despite repeated criticism (Gray, 1929; Hogben, 1931; Kavanagh and Richards, 1934; Feller, 1940; Andrewartha and Birch, 1954; Smith, 1952, 1954; Pielou, 1969, 1977).

An alternative mathematical means of describing population growth is the Leslie (1945) matrix, a density-independent model. Because this model with its constant age-specific fertility and mortality rates could result in a geometrically increasing population once a stable age distribution was established, Leslie (1948) modified it so that density-dependent effects were included and population growth could approach asymptotically an upper limit, as does the logistic equation.

The density dependence of the logistic equation of the mathematicians fits well the ideas of naturalists and experimentalists regarding density-dependent regulation of population size. Nevertheless, the existence of density-dependent regulating mechanisms that are sufficiently effective to depress population growth has been difficult to demonstrate. A review of the literature on growth in animal populations (Lack, 1954) and a subsequent analysis of long-term population studies of birds (Lack, 1966) failed to demonstrate to Lack's satisfaction the existence of density-dependent regulating mechanisms. He was forced to defend the density-dependent regulation of animal populations on logical grounds: "The absence of field evidence does not, and will not, make the advocates of density-dependent regulation change their minds . . . because, given certain assumptions about the persistence of natural populations, the existence of density-dependent regulation becomes a logical necessity" (Lack, 1966). He suggested that regulation by density-dependent mortality occurred during those portions of a species' life history for which he had no information.

Reddingius (1971) undertook a logical analysis of the opposing population models and concluded (1) that "negative density-dependence of net reproduction is neither necessary nor sufficient for a population to tend to equilibrium, or even to persist" (p. 21), (2) that the thinking of ecologists who suggest that weather fluctuations (for example) are mainly responsible for density fluctuations is not as illogical as other ecologists contend (p. 83), (3) that the available data are too crude or are inappropriate for statistical tests that might reject any of the population theories so far proposed (e.g., p. 116), and (4) that it seems desirable "to replace the metaphor 'struggling for existence' by the other metaphor 'gambling for existence'" (p. 98).

The contending ideas regarding the determination of a population's numbers have been so exhaustively reviewed, reinterpreted, and reviewed again (see esp. Andrewartha and Birch, 1954; Thompson, 1956; Solomon, 1957; Nicholson, 1958b; Milne, 1962; Bakker, 1964; Huffaker and Messenger, 1964b; Lack, 1966; Clark *et al.*, 1967; Reddingius, 1971) that another detailed review of the arguments seems unjustifiable. The purpose of this book is not to sort out previous theories but to present an explicit interpretation of population dynamics that begins with different assumptions than previous efforts. It generates predictions that appear to be consistent with many observations collected in the field and laboratory.

THE POPULATION PROBLEM

Field ecologists always sample a small portion of a species' population at some restricted site within the species' range (Fig. 1.1), throughout which

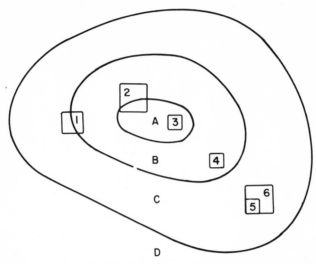

Figure 1.1. The geographic range of a hypothetical species. The most favorable conditions for survival and successful reproduction are in area A, intermediate conditions at B, and less favorable at C. Area D is not favorable at all, and the species does not exist there. The ecologist samples the species at one study area or another (e.g., 1, 2, 3, 4, 5, or 6). The population parameters, such as age-specific birth and death rates and age structure, may be different between study areas, depending on the conditions prevailing in the local area and on the size of the area chosen for study. The dynamics of a population in any one area is not necessarily the same as the dynamics of populations in other areas.

the population parameters—e.g., density, birth and death rates—vary. The minimal data collected include the sample population's numbers at intervals during some period of time. More ambitious research provides some measure of age-specific death rates (or survivorship) and some measure of the reproductive success of some individuals within the sample population (fecundity). Additional data describe the physical and biological environments of the individuals of the population being studied (e.g., temperature, humidity, pH, predation, density). Often, field work is supplemented by laboratory work to measure birth rates and death rates under different conditions, such as crowding and temperature. All this information is analyzed in order to describe the cause and effect relations determining population numbers.

Whereas laboratory populations provide data on the growth of populations from small founder populations to much larger steady-state populations (A in Fig. 1.2), field ecologists rarely observe such growth. Most field studies record temporal fluctuations (B in Fig. 1.2). The cause(s) for the change in numbers during the initial growth phase of a population may or may not be the same cause(s) for the changes in numbers during the

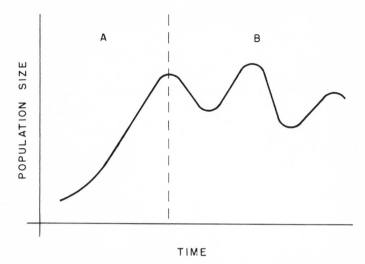

Figure 1.2. Growth of a hypothetical population. The growth phase (A) is usually studied by the laboratory ecologist. Field ecologists are usually restricted to studying populations in the fluctuating phase (B).

fluctuating phase. We should not ignore the possibility that different explanations may be involved.

Certain fluctuations observed in the field are of little theoretical interest. For instance, when one is considering a species with a seasonal reproductive period (Fig. 1.3), the increase in population size during the reproductive period and the decrease in the postreproductive period do not pose difficult problems. What is of interest is the magnitude of change in population size from some particular time to a comparable later time.

A difficult problem for the field biologist is determining emigration and immigration rates. Losses owing to emigration usually cannot be separated from losses from deaths. Thus, survivorship measured in field studies may not reflect the actual life expectancies of individuals born within the spatial unit being studied. A knowledge of actual survivorship is necessary for a complete understanding of the dynamics and evolution of populations, but so far as *determination of numbers* of the population within the spatial unit being studied is concerned, emigration and mortality can be considered equivalent.

Distinguishing gains from births and gains from immigration is also important and nearly as difficult to do, but again for *determination of numbers* these need not be rigorously separated. Survivorship schedules calculated from age distributions of animals recorded within the spatial unit will reflect the effects of immigration and emigration.

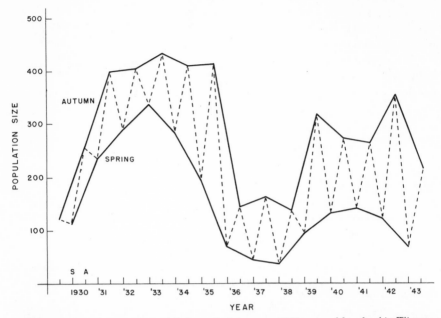

Figure 1.3. The numbers of *Colinus virginianus* on 4500 acres of farmland in Wisconsin. The upper curve gives the numbers at the beginning of autumn, and the lower curve gives the number at the beginning of spring. Fluctuations in numbers between seasons are more easily explained and less interesting than fluctuations between the same season in different years. (Redrawn from Errington, 1945.)

Projecting the future growth of natural populations, or even attempting explanations for recorded past events, requires several assumptions, none of which is highly probable. We need a measure of the population's age structure, and we must assume that it is stable (or we must assume that some other, necessarily more arbitrarily chosen age distribution is stable). We also need a measure of the age-specific birth rates, and we must assume that they are constant or, at least, average values. From the stable age distribution, age-specific mortality rates and survivorship schedules can be calculated. Survivorship schedules constructed from existing age structures are vertical survivorship schedules, and these must be assumed to remain constant during the time period of interest. Horizontal survivorship schedules, constructed from known ages of death, should be the same as vertical survivorship schedules if the age distribution is stable and immigration and emigration rates remain constant (although not necessarily equal). Horizontal survivorship schedules, calculated from known ages of death of a marked cohort, however, will differ from both these survivorship schedules because immigrants will not be included in the calculations.

The sample population is always the subject of interest. Therefore, "population density," "population numbers," and "population size" are equivalent terms because the population is always restricted to a definite space. Furthermore, the values calculated from our sample population are applicable only to that population. The interpretation of the dynamics of the study population may or may not be valid for other populations of the same species.

In order to ease the presentation of complex systems, let us assume that the sexes of the populations discussed in this book occur with equal frequency and do not differ in their demographic characteristics. Analysis of seasonally reproducing populations is simplified by assuming that recruitment occurs as a series of pulses spaced at (usually) yearly intervals and that censuses are taken immediately after each impulse (Caughley, 1967). The principles developed here can be applied to more complex systems by treating each sex separately and by using any convenient time period.

Effect of Sample Area Size

Just as the species composition of a community (Gleason, 1922) and various measures of distribution and abundance (Curtis and McIntosh, 1950; Evans, 1952) are affected by the size of the area being studied (Figs. 1.4 and 1.5), the dynamics of the population within the selected study area will be influenced by the study area's size, which I will refer to as the "area effect."

The importance of the area effect on interpretations of dynamics is easily seen by considering the well-known case of the relationship between the

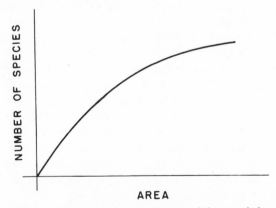

Figure 1.4. The species–area curve. As the size of the sampled area increases, the number of species increases toward an asymptote.

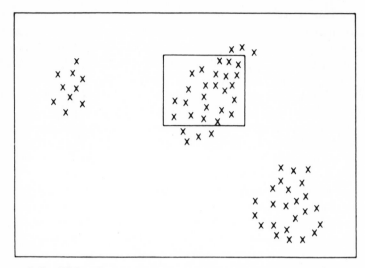

Figure 1.5. Within the larger area the individuals (x) are clumped, but within each aggregation the individuals are regularly or uniformly distributed. Each aggregation appears to be randomly distributed. In nature, the distribution of suitable habitat may be random (with respect to geographic location), resulting in a random distribution of aggregations. The dispersion of the individuals, however, would be clumped, and within each aggregation territorial behavior, for example, could result in a uniform dispersion of individuals.

prickly pear (*Opuntia* sp.) and the cactus moth (*Cactoblastis cactorum*) in Australia (Dodd, 1959). After its introduction, *Opuntia* spread rapidly and ruined millions of acres of pasture and rangeland. Subsequently, *Cactoblastis* was imported, and it, too, rapidly spread, reducing the distribution and abundance of *Opuntia*. Now, throughout their mutual range, *Opuntia* and *Cactoblastis* are rare. If we sampled a large area, a series of annual censuses of *Opuntia* and *Cactoblastis* would show slight variations about some mean value. We might be tempted to believe that both populations were "regulated." Yet a small sample area would reveal a different dynamic relationship between *Opuntia* and *Cactoblastis*. If our sample area enclosed a clump of *Opuntia* before the arrival of *Cactoblastis*, we would observe the arrival of *Cactoblastis* followed by a continuous increase in *Cactoblastis* numbers and decrease in *Opuntia* until the *Opuntia* clump was completely destroyed; then *Cactoblastis* numbers would decline to zero. Evidently, the low-density, steady-state populations of *Opuntia* and *Cactoblastis* in the larger study area result from the dispersal of both *Opuntia* seeds and *Cactoblastis* adults from smaller areas where local events of establishment, growth, and extinction prevail. It seems, then, that there are no density-dependent factors or effects maintaining these two populations in the larger study area. The system is maintained by the dispersal capabilities of both populations. *Opuntia* survives because its seeds disperse to areas not occupied by *Cactoblastis*. They have the opportunity to germinate, grow, and set seed before being discovered and devoured by *Cactoblastis*. *Cactoblastis* survives because the moths can disperse, find the new clumps of *Opuntia*, mate, and produce another generation of dispersing moths before destroying the *Opuntia* clump. The steady-state size of the larger *Opuntia* and *Cactoblastis* populations would be different if each species had greater or lesser abilities to disperse and if a more or less efficient *Opuntia* finder or *Opuntia* eater had been introduced.

Huffaker (1958b) showed a similar area effect in the dynamics of an experimental predator–prey interaction between two mite species. The six-spotted mite (*Eotetranychus sexmaculatus*) was the prey species, and *Typhlodromus occidentalis* was its predator. The prey fed on oranges that were either clumped or randomly placed among a group of similarly sized rubber balls. In the most complex arrangement, the predator–prey system persisted for three oscillations before the predator population died out (Fig. 1.6). However, during the coexistence of the predator and prey, the prey population on each orange was eliminated. The system was maintained by the dispersal of the prey species to oranges unoccupied by predators. Eventually, the predators found the prey and annihilated them. On each orange, the prey population was doomed if found by the predator, but the larger system showed coexistence and oscillations in numbers of both predator and prey species. Again, in this system we find no evidence of density-dependent regulation of either predator or prey populations.

Of course, not all large populations must consist of local populations going through phases of establishment, growth, and extinction. I mention these examples to show the importance of the study area's size on the interpretation of the population's dynamics.

Bakker (1970) discussed the "cat and mouse" relationships of *Opuntia* and *Cactoblastis* and of *Eotetranychus* and *Typhlodromus* with reference to the meaning of "regulation." He could not see any advantage in calling these cases "regulation" because "the mechanism which is responsible for this apparent regulation can be perfectly understood from the processes taking place in the various subpopulations . . . and from the statistics of the system" (p. 566). He further suggested that " 'regulation' is *not a property* of a population in the way that homeostasis may be a physiological property of an individual organism" (emphasis in original, p. 566).

In later sections, we will see further examples of the area effect on interpretations of population dynamics, specifically with reference to territoriality and to the competitive exclusion principle.

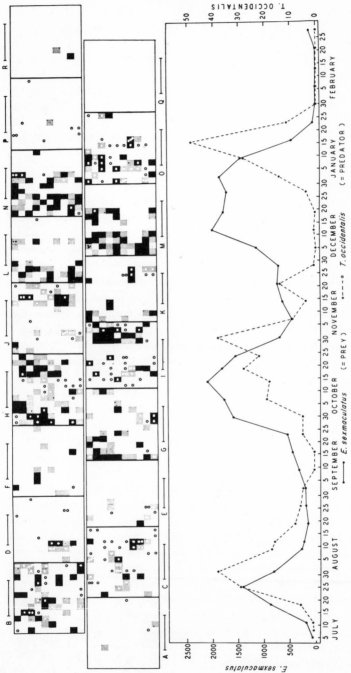

Figure 1.6. Three oscillations in the density of a predatory mite, *Typhlodromus occidentalis*, and its prey, *Eotetranychus sexmaculatus*, another mite. The populations are living on 120 oranges, each of which has 1/20 of the orange area exposed and is separated from the others by a complex maze of Vaseline barriers. The lower graph records the numbers of predator and prey per orange area. The upper diagram records the densities and positions of predator and prey at the 120 orange positions. The horizontal lines marked by A, B, etc., indicate the period on the time scale represented by each chart. Prey density: 0–5 nil (white); 6–25 low (light stipple); 26–75 medium (horizontal lines); 76+ high (solid black). Predator density: 1–8 (one white circle). (From Huffaker, 1958b.)

10

DEFINITIONS

Many ecologists seem to recognize a simple dichotomy of "density-dependent" factors versus "density-independent" factors when discussing the environmental variables that can affect a population's numbers, despite a lengthy discussion of the meanings of the terms (Andrewartha and Birch, 1954; Andrewartha, 1958, 1959; Nicholson, 1954a, 1958a, 1959; Milne, 1958b, 1959; Solomon, 1958a,b; Varley, 1958, 1959a,b; Royama, 1977). As there is no consensus, it seems necessary to provide the following definitions:

1. *Density-dependent factor*. Any component of the environment whose intensity is correlated with population density and whose action affects survival and reproduction.

(a) *Effective density-dependent factor*. Any component of the environment whose action is sufficiently intense to regulate population size; that is, at some point the factor must suppress a further increase in population numbers.

(b) *Ineffective density-dependent factor*. Any component of the environment whose action adversely affects survival and reproduction but not sufficiently to suppress a further increase in population numbers.

(c) *Inverse density-dependent factor*. Any component of the environment that increases (or promotes) survival and reproduction with increasing population density.

2. *Density-independent factor*. Any component of the environment that affects survival and reproduction but whose occurrence or intensity is not correlated with population density.

Density-dependent and density-independent factors, then, are those components of the environment that produce changes in survivorship, fecundity, or both. It is important to distinguish between cause and effect when using these terms. For example, the population's density will affect the intensity of intraspecific competition but will not determine the weather. But both competition and the weather can cause density-dependent responses in survivorship and fecundity. Andrewartha and Birch (1954) were looking at the effects of these factors when they wrote, "We conclude that 'density-independent' factors do not exist."

Density-dependent factors cannot always be assumed to be effective in regulating population size. For example, in the Great Tit (*Parus major*), decreasing clutch size is correlated with increasing population density (Fig. 1.7A). Whatever the density-dependent factor is (perhaps intraspecific competition for food), it is not regulating the population's size because the most dense populations consisting of the least fecund individuals are the ones that produce the most flying young (Fig. 1.7B). Thus, the reduction in

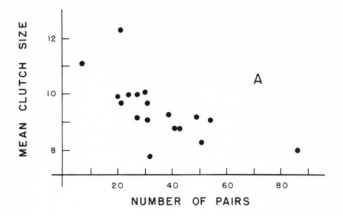

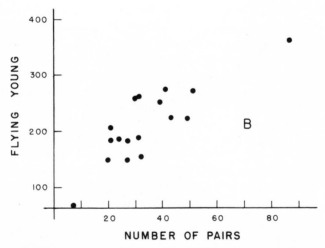

Figure 1.7. Reproduction in the Great Tit. (A) The mean clutch size decreases with density (from Lack, 1966). (B) The total number of flying young produced increases with increasing density. (From Lack, 1966.)

clutch size is an ineffective density-dependent response. Lack (1966) certainly recognized this, but Krebs (1970), studying the same data as Lack did, suggested that "clutch size and hatching success are density-dependent and sufficiently so to regulate the population at the observed level (assuming that there is in addition a fairly large density-independent mortality)." The parenthetical statement indicates that clutch size is an ineffective density-dependent factor.

On the other hand, density-independent factors, such as the weather, can be very effective in depressing survivorship, reproductive success, and population size in insects (Andrewartha and Birch, 1954, 1960; Ehrlich *et al.*, 1972) and in birds (Jehl and Hussell, 1966; Tompa, 1971).

Royama (1977) distinguished between statistical dependence and causal dependence. Some environmental factors may be statistically correlated with population density without being causally related. Although these factors might be termed density-dependent (he identified them as type B density-dependent), they are not likely to regulate population density for long, if at all. Type A density-dependent factors are causally related to density and therefore regulate population density. The density-dependent factors we are interested in are type A.

3. *Stability*. The quality of "continuance without change" (Urdang, 1968). A stable population should have constant birth rates, death rates, age distribution, and size, the values of which vary only by chance, or, after disturbance, the population should return to its original condition.

The term "stability" seems to have two different meanings in the ecological literature. Some ecologists (Murdoch, 1966; Hairston *et al.*, 1968; May, 1973) define stability as it is used in everyday language (Urdang, 1968, quoted above) and in physics and engineering (Preston, 1969). In this sense, stable systems return to their equilibrium values after disturbance, either monotonically or through a series of damped fluctuations. Stable systems are maintained by negative feedback loops.

A second conception of stability, called "neutral stability" by May (1973), considers a population "stable" when it fluctuates within prescribed limits; that is, the population does not increase indefinitely or decrease to extinction (Lack, 1954, 1966; Nicholson, 1958a; Preston, 1969; Watt, 1969; Royama, 1977). These fluctuations are sometimes extraordinary, as shown, for example, by laboratory populations of the Australian sheep blowfly (*Lucilia cuprina*), which alternately grow to 3500 adults and crash to nearly zero (Fig. 1.8; Nicholson, 1958a). Such fluctuating populations are considered "stable" because they are "persistent" (i.e., not extinct). Interestingly, logistic equations with time-delayed, density-dependent effects on growth rates model some fluctuating populations fairly well, although the biological mechanisms leading to the delays are not known (May, 1976a).

By defining extant populations as stable, and stable populations as regulated populations, the range of potential models describing population dynamics is severely limited. The search for adequate explanatory models of population phenomena should not begin with *a priori* idealized conceptions but with an open mind. A model should be evaluated against the data of experience rather than preconceived notions. The notion of stability assumed extinction to be rare, but Andrewartha and Birch (1954) suggested

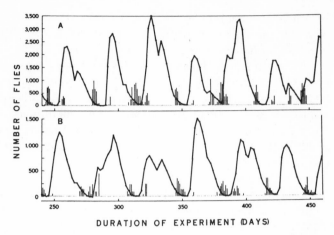

DURATION OF EXPERIMENT (DAYS)

Figure 1.8. Numbers of adult *Lucilia cuprina* (heavy line) in population cages. Vertical bars represent the numbers of adults eventually emerging from eggs laid on dates of plots, indicating the severe competition between larvae for food. Larval food was 50 gm/day in A and 25 gm/day in B. (From Nicholson, 1958a.)

that the extinction of local populations is a common event. And, as we have already seen, persisting large populations of *Opuntia* and *Cactoblastis* and of *Eotetranychus* and *Typhlodromus* consist of smaller populations, which regularly become extinct.

Either conception of stability discussed above seems only to place unnecessary restrictions on the kinds of models that can be imagined and tested. In this book, neither stability nor instability of populations is assumed.

4. *Regulation and limitation.* These two terms are often used interchangeably. This can only lead to confusion. Inasmuch as a standard argument is, "Populations are persistent; therefore, populations are regulated" (e.g., Lack, 1966; Royama, 1977), the term "regulation" should be used with reference to models in which density-dependent factors maintain a population's numbers, and indeed this seems to be its universal meaning.

Limitation should be used with reference to models that propose a level above which a population's numbers cannot increase because the birth rate does not exceed the death rate. Below that level, birth and death rates may be constant or variable, and usually the birth rate exceeds the death rate, but nevertheless they are not necessarily correlated with population density.

The source of confusion in the use of these terms is the seemingly legitimate notion that "limiting factors" are responsible for setting either the carrying capacity about which population size is regulated or setting the level that limits further population growth.

5. *Population and environment.* In this book, a population is any arbitrarily chosen collection of individuals of the same species. This is the usual definition of an ecological population (see reviews in Clark *et al.*, 1967; Reddingius, 1971). As such, the dynamic relationships developed later are applicable to any more restrictively determined population, such as a deme. Nicholson (1933 and later papers), however, suggested that populations are ecological entities with properties of their own.

The environment has been defined in terms of its action on individual organisms (Andrewartha and Birch, 1954), populations (Nicholson, 1933, 1954a,b, 1958a,b), or individuals, populations, or biocoenoses, depending on the context (Reddingius, 1971). In this book, only individuals have environments. It is convenient, however, to write about a "population's environment," but this should always be interpreted to mean the environments of the individuals making up the population under discussion. Environmental factors affect the individuals' chances of surviving and reproducing. Environmental factors include "other individuals within the same population," and thus each individual has a different environment from the others making up a population. Population statistics reflect the individuals' probabilities of survival and reproduction.

QUESTIONS

Typical population data from laboratory populations (Fig. 1.8) and natural populations (Fig. 1.9) show average differences in population size and differences in the magnitude of fluctuations in size. The population ecologist is interested in answering various questions: (1) Why does the sample population grow to a particular size instead of some larger or smaller size? (2) Why does the population's size fluctuate about some mean value instead of remaining constant or fluctuating even more greatly? (3) Why does a population at one place differ in size and in the amplitude of its fluctuations from another population of the same species in another area? (4) Are the factors that determine the size of a population and the amplitude of its fluctuations in one area the same as those that determine the size and amplitude of fluctuations of a population in a larger area? (5) Are the factors that determine the commonness or rarity (abundance) of a species within the sample areas the same as those that determine the extent of the species' geographic range (distribution)?

In providing explanations, the population ecologist should keep a clear

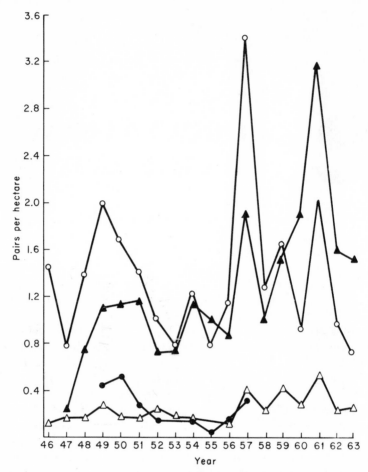

Figure 1.9. Density of breeding pairs of four populations of the Great Tit in Europe. Key: ▲, Marley; ○, Forest of Dean; ●, Breckland; △, Veluwe. (From Lack, 1966.)

distinction between how the species' characteristics relate to the species' present distribution and abundance and how those species' characteristics evolved in response to past environments (Birch and Ehrlich, 1967). This book is concerned primarily with the first kind of explanation. The evolution of life history characteristics, however, is treated in Chapter 4.

2 | POPULATION MATHEMATICS

The goal of any science is to produce unifying generalizations, called "principles" or "laws," that model the cause and effect relationships between several variables within a system. Given a set of generalizations, we can determine the effects of a perturbation to one variable on other variables, and these we compare with observations from nature or from the laboratory in judging the quality of the model. Models of high quality become laws or principles. These are always simple compared with the complexity of the perceptible world. Developing models simulating the complexity of natural systems is important in the management of resource populations (Watt, 1968) or in projecting future states of the world system (Meadows *et al.*, 1972), but such models do not preclude searching for principles that relate the parameters of the system. Complex models, the solutions of which require the use of high-speed computers, can give precise solutions to specific problems, provided that the generalizations used in describing the relationships within the system are correct, but they are not likely to produce a new unifying generalization (May, 1973; Murray, 1975), which might be construed as a "law of nature." For example, a systems model could be constructed to predict the motion of a falling feather, and another model could be constructed to predict the motion of a falling hammer. A series of *ad hoc* models could be constructed for a variety of falling objects, but it is unlikely that this approach could lead us to Galileo's equation for freely falling objects. Galileo's equation oversimplifies the reality of our perceptions, but it has heuristic value beyond that of a series of *ad hoc* equations. We "know" that a feather does not free-fall on earth because of air resistance, but this could not be known without Galileo's equation, nor could we predict the free-fall of a feather on the moon without Galileo's equation. Generalizations are not intended to describe nature in its infinite variety. Instead, they bring order to an otherwise disorderly world (Bronowski, 1965).

Bartlett (1973) remarked on the almost inevitable confrontation between biologists and mathematicians in formulating models of biological processes. The population biologist begins with a mass of data, or at least a diverse array of observations, for which he wants to develop a quantitative description of the cause and effect relationships such that he is able to predict changes in population size, if only to prove that he does in fact understand the relationships or to establish strategies of resource exploitation or conservation. The mathematician, in contrast, is more "likely to be looking for an idealized model capable of theoretical investigation" (Bartlett, 1973). Pielou (1977) adds, "The investigation of mathematical models often seems to be motivated more by an interest in the mathematics of the model, and a striving for mathematical elegance, than by an interest in the model's ecological implications."

Mathematical analysis of populations is essential for developing an understanding of population processes, and, although idealized models are simpler than the natural phenomena they intend to describe, they are useful for organizing data and thoughts and for providing insights. Idealized mathematical models are often unappreciated by field ecologists, but Bartlett (1973) correctly asked, "Until we can understand the properties of these simple models, how can we begin to understand the more complicated real situation?"

This book is a field ecologist's attempt to discover quantitative relationships between population parameters in order to understand the observable world and perhaps to serve as an antidote to the recent trend toward mathematical elegance (e.g., May, 1976b; Pielou, 1977). For the field biologist, mathematics is no more than a tool, and the minimal set of tools falls into three distinct categories: (1) the exponential and logistic equations describing population growth (Verhulst, 1838; Pearl and Reed, 1920), (2) actuarial equations relating several population parameters (Lotka, 1925), and (3) the projection matrix (Leslie, 1945, 1948). These will be considered only to the extent necessary to follow the theoretical development in later chapters. Fuller treatment of population mathematics can be found in Pielou (1969, 1977), Mertz (1970), and Poole (1974).

EXPONENTIAL AND LOGISTIC EQUATIONS

The dynamics of population growth has been described most frequently by exponential and logistic equations (Verhulst, 1838; Pearl and Reed, 1920). These describe the change in numbers of populations in unlimited and limited situations, respectively. In constant and unlimited conditions, the rate of change in population size is given by

$$\frac{dN}{dt} = rN, \tag{2.1}$$

where r is the instantaneous rate of increase per head and N is the population's size. Integrating, we have

$$N_t = N_0 e^{rt} \tag{2.2}$$

where N_0 is the initial population size, N_t is the population's size at time t, and e is the base of natural logarithms.

The exponential equation is useful in projecting population growth when conditions are assumed to be constant and unlimiting. These conditions prevail only at low population densities, if then, so the projection is primarily of theoretical interest. Populations live in limited situations, and thus they do not grow indefinitely and often show sigmoid growth.

The classic equation describing sigmoid growth of populations in limited situations is the logistic, which was independently derived by the demographers Verhulst (1838) and Pearl and Reed (1920). The growth rates of the human populations these investigators were studying were declining, and there was interest in determining the size of future populations. In the logistic equation, the population's growth rate declines as its numbers increase,

$$\frac{dN}{dt} = rN\left(\frac{K - N}{K}\right), \tag{2.3}$$

where N is the population's size, K is the population's saturation value for the prevailing environmental conditions, and r is the population's "intrinsic rate of increase." The fitting of the logistic equation to empirical data has been described by Ricklefs (1967), Poole (1974), and Crossner (1977).

The logistic equation has been criticized frequently (Gray, 1929; Hogben, 1931; Kavanagh and Richards, 1934; Feller, 1940; Andrewartha and Birch, 1954; Smith, 1952, 1954; Pielou, 1969, 1977), mainly because more than one equation can fit the same data set, and each equation has different biological properties. Furthermore, population growth of animals with complex life histories does not often conform to the logistic (Andrewartha and Birch, 1954). The advantages of the logistic are its simplicity, compared with alternatives, and its "reality," meaning that its constants have biological meaning (Andrewartha and Birch, 1954). The logistic is also attractive because its density-dependent growth rate is consistent with the logical notion of density-dependent regulation (Lack, 1966), although, it should be noted, at least one promoter of density dependence explicitly rejected the logistic equation as a model of population growth (Nicholson, 1958b).

The logistic model, as any model, has limitations. Pielou (1969, 1977) listed the restrictive assumptions, which "must frequently be false": (1) The abiotic environment is sufficiently constant that birth and death rates are unaffected; (2) crowding affects all individuals of the population equally; (3) birth and death rates respond instantly to changes in density; (4) the population's growth rate is density-dependent even at low densities; (5) the population maintains a stable age distribution; and (6) the females in sexually reproducing populations always find mates. Pielou (1969, 1977) mentions that if assumptions (4) and (6) were incorrect, the effects would cancel each other. But even so, assumptions (3) and (5) are incompatible. Any change in the age-specific death rates changes the stable age distribution (see next section). Thus, under no circumstances can these six assumptions be exhibited by a population.

LOTKA'S EQUATIONS

As a model of population growth, the logistic equation relates only the population's growth rate with its density. Several other population parameters vary as conditions change, such as the age-specific birth and death rates, the true rate of increase, the rate of increase per generation, and the population's age distribution. All of these are related in a series of equations developed during the first quarter of this century by Lotka (1907a,b, 1913a,b, 1922, 1925; Sharpe and Lotka, 1911; Dublin and Lotka, 1925). These relationships are most important for understanding the discussion of population dynamics presented in later chapters, and so they should be considered in some detail now.

Mathematical analysis begins with the estimation from population data of the proportion of each cohort that survives to each age x (the l_x schedule) and the average number of eggs laid or young born to individuals of age x (the m_x schedule). From these two schedules, several other population statistics can be calculated by using Lotka's equations. It is usual in population analysis to consider only female survival and female births, but

I prefer to think in terms of whole populations and, for simplicity, assume that the demographic characteristics of males and females do not differ. If males do differ from females, adjustments will have to be made regardless of which simplifying assumption is made in order to describe the dynamics of the whole population.

The parameter r, known as the "true," "standardized," or "stable rate of increase" (Dublin and Lotka, 1925), the "incipient rate of increase" (Lotka, 1927), the "inherent" or "intrinsic rate of increase" (Lotka, 1943), the "innate rate of increase" (Andrewartha and Birch, 1954), the "Malthusean parameter" (symbolized m by Fisher, 1930), and the "ultimate rate of natural increase" (Mertz, 1970, 1971a), is given by the equation

$$1 = \int_0^\infty l_x m_x e^{-rx} dx. \tag{2.4}$$

The population's instantaneous birth rate b is given by

$$1/b = \int_0^\infty l_x e^{-rx} dx, \tag{2.5}$$

and its instantaneous death rate d by

$$b - d = r. \tag{2.6}$$

The population's stable age distribution is determined from

$$c_x = bl_x e^{-rx}, \tag{2.7}$$

where c_x is the proportion of the total population made up of age class x.

The calculation of r from Eq. (2.4) is tedious (Dublin and Lotka, 1925). Field or laboratory data on nonhuman populations are often subject to considerable sampling error, so Birch (1948) suggested a simpler method to estimate the value of r that is sufficiently precise under such conditions (his paper should be consulted for details). Birch (1948) begins by estimating the value of r from

$$r = \frac{\ln R_0}{T}, \tag{2.8}$$

where R_0 is the net reproductive rate (the population's multiplication rate per generation), which is given by

$$R_0 = \int_0^\infty l_x m_x dx, \tag{2.9}$$

and T is the mean generation time, which is given by

$$T = \frac{1}{R_0} \int_0^\infty l_x m_x x \, dx. \tag{2.10}$$

Equation (2.10) provides only an approximation of the generation time, and therefore Eq. (2.8) can only estimate the value of r. Leslie (1966) showed that a more accurate estimate of the mean length of a generation is given by

$$\bar{T} = \int_0^\infty l_x m_x x e^{-rx}\, dx. \tag{2.11}$$

Once r is estimated from Eq. (2.8), trial values of r are substituted in Eq. (2.4). In practice, however, summation equations are used rather than integrals.

In order to illustrate the method of calculating various population statistics, a hypothetical population with simple l_x and m_x schedules has been constructed (Table 2.1). From the given l_x and m_x schedules, $\Sigma l_x m_x$ and $\Sigma l_x m_x x$ can easily be calculated. Thus,

$$R_0 = \Sigma l_x m_x = 5.0,$$

$$T = \frac{1}{R_0} \Sigma l_x m_x x = \frac{28.0}{5.0} = 5.6,$$

and

$$r = \frac{\ln R_0}{T} = \frac{1.6094}{5.6} = 0.2874.$$

By trial and error substitution in

$$1 = \Sigma l_x m_x e^{-rx},$$

$$r = 0.3010.$$

The instantaneous birth rate is, from

$$1/b = \Sigma l_x e^{-rx} = 2.0765,$$

$$b = 0.4816.$$

The instantaneous death rate is

$$d = b - r = 0.4816 - 0.3010 = 0.1806.$$

The stable age distribution of this population is given in column (10) of Table 2.1.

The theories of population growth developed in later chapters focus attention on steady-state conditions. The mathematics of steady-state populations is simple compared with that of growing or declining populations. In steady-state populations, $r = 0$ and $e^{-rx} = 1$. Therefore, Lotka's fundamental equations become

$$1 = \Sigma l_x m_x, \tag{2.12}$$

$$1/b = \Sigma l_x, \tag{2.13}$$

$$c_x = b l_x. \tag{2.14}$$

TABLE 2.1

Demographic Data of Hypothetical Populations

(1) Age (x)	(2) q_x	(3) l_x	(4) m_x	(5) $l_x m_x$	(6) $l_x m_x x$	(7) e^{-rx}	(8) $l_x e^{-rx}$	(9) $l_x m_x e^{-rx}$	(10) c_x	(11) q_x	(12) c_x
1	0.3339	1.0	0	0	0	0.7401	0.7401	0.0	0.3564	0.1000	0.1818
2	0.3420	0.9	0	0	0	0.5477	0.4929	0.0	0.2374	0.1111	0.1636
3	0.3528	0.8	0	0	0	0.4054	0.3243	0.0	0.1562	0.1250	0.1455
4	0.3660	0.7	2	1.4	5.6	0.3000	0.2100	0.4200	0.1011	0.1429	0.1273
5	0.3822	0.6	2	1.2	6.0	0.2220	0.1332	0.2664	0.0641	0.1667	0.1091
6	0.4091	0.5	2	1.0	6.0	0.1643	0.0822	0.1643	0.0396	0.2000	0.0909
7	0.4444	0.4	2	0.8	5.6	0.1216	0.0486	0.0973	0.0234	0.2500	0.0727
8	0.5077	0.3	2	0.6	4.8	0.0900	0.0270	0.0540	0.0130	0.3333	0.0545
9	0.6250	0.2	0	0	0	0.0666	0.0133	0.0	0.0064	0.5000	0.0364
10	1.000	0.1	0	0	0	0.0493	0.0049	0.0	0.0024	1.0000	0.0182
Sum		5.5		5.0	28.0		2.0765	1.0020	1.0000		1.0000

$r = 0.3010$
$R_0 = 5.0000$
$b = 0.4816$

$r = 0.0000$
$R_0 = 1.0000$
$b = 0.1818$

If a steady-state population sustains any additional mortality, it will decline to extinction unless m_x increases [Eq. (2.12)]. The birth rate of a steady-state population is simply the reciprocal of the sum of l_x values [Eq. (2.13)]. It is convenient to think of the birth rate from Eq. (2.13) as the *replacement rate* b_r of any population with a given l_x schedule, regardless of the value of r. If $r > 0$, then b from Eq. (2.5) will be greater than the replacement rate, and the population will grow. If $r < 0$, then b from Eq. (2.5) will be less than the replacement rate, and the population will decline.

For the given l_x schedule [column (3) of Table 2.1] the replacement rate is 0.1818, considerably lower than the population's actual birth rate, 0.4816. The steady-state age distribution is given in column (12) of Table 2.1 and should be compared with the stable age distribution of the growing population with the same l_x schedule, shown in column (10) of Table 2.1.

There are limitations to the use of Lotka's equations and to the interpretation of the statistics. The primary assumption is that the population has a stable age distribution. A further limitation is that l_x and m_x must be calculated on a per annum basis, or the population must be continuously breeding in uncrowded conditions.

Demography of Intermittently Breeding Populations

When time interval x is less than 1 year and when individuals breed seasonally, neither r nor R_0 provides any indication of the rate of growth of a population from one year to the next, which more often than not is the statistic of interest in many population studies (Leslie and Ranson, 1940). It is possible for a population to decline even when $r > 0$ and $R_0 > 1$. Two additional statistics must be calculated for seasonally breeding populations. These are (1) the momentary coefficient of increase per year, r'_a, and (2) the number of times the initial population increases in 1 year, $P'_{52-t}e^{rt}$ or $P'_{12-t}e^{rt}$, when t is measured in weeks or months, respectively.

Leslie and Ranson (1940) assumed (1) that the beginnings of consecutive breeding seasons were exactly 1 year apart and (2) that during the breeding season a population of N_0 individuals increased (approximately) exponentially. Thus, at the end of the breeding season the population N would be

$$N = N_0 e^{rt},$$

where t is measured in weeks or months.

Following the end of the breeding season, the population's numbers decline depending on the l_x schedule and the population's age distribution. If P'_{12-t} is the proportion of the population still alive at $12 - t$ months after the end of the breeding season, the population at the end of 1 year will be

$$N = N_0 e^{rt} P'_{12-t}, \tag{2.15}$$

and the momentary coefficient of increase per year is

$$r'_a = \ln(P'_{12 - t}e^{rt}). \qquad (2.16)$$

The data given in Table 2.1 can serve as an illustration, assuming t is measured in months. With the given l_x schedule, there are three populations with different age distributions that are of interest. The first is the stable age distribution of the growing population ($r = 0.3010$), shown in column (10), Table 2.1. The second is the steady-state age distribution ($r = 0.0$), shown in column (12), Table 2.1. Finally, one can consider the possibility that all individuals entering the nonbreeding season are prereproductive individuals (say, overwintering eggs, i.e., age class 1). The monthly declines of three populations of 1000 members each, with l_x schedule [column (3), Table 2.1] and with these age distributions appear in Table 2.2. The most rapid decline occurs in the steady-state population because its members are on average older than the members of the other two populations.

The values for r'_a and for $P'_{12 - t}$ can be calculated from the data in Table 2.2. These have been worked out for the growing and steady-state populations (Table 2.3 and Fig. 2.1). Despite $r > 0$ and $R_0 > 1$, the population's

TABLE 2.2

Population Decline during Nonbreeding Season

Time (months)	Number of survivors		
	Population		
	A[a]	B[b]	C[c]
0	1000	1000	1000
1	867	818	900
2	740	654	800
3	616	508	700
4	496	381	600
5	380	272	500
6	275	181	400
7	179	108	300
8	97	54	200
9	36	18	100
10	0	0	0

[a] Age distribution from column (10), Table 2.1.
[b] Age distribution from column (12), Table 2.1.
[c] All individuals in age class 1.

TABLE 2.3

Annual Rates of Increase for Seasonally Breeding Populations

Length of breeding season (months)	Population A				Population B	
	e^{rt}	P'_{12-t}	$P'_{12-t}e^{rt}$	r'_a	P'_{12-t}	r'_a
1	1.3512	0.0	0.0	—	0.0	—
2	1.8258	0.0	0.0	—	0.0	—
3	2.4670	0.036	0.0888	−2.4212	0.018	−4.0174
4	3.3334	0.097	0.3233	−1.1290	0.054	−2.9188
5	4.5042	0.179	0.8062	−0.2154	0.108	−2.2256
6	6.0861	0.275	1.6737	0.5150	0.181	−1.7093
7	8.2235	0.380	3.1249	1.1394	0.272	−1.3020
8	11.1117	0.496	5.5114	1.7068	0.381	−0.9650
9	15.0143	0.616	9.2488	2.2245	0.508	−0.6773
10	20.2874	0.740	15.0127	2.7089	0.654	−0.4246
11	27.4125	0.867	23.7667	3.1683	0.818	−0.2009
12	37.0401	1.000	37.0401	3.6120	1.000	0.0000

breeding season must be 5.25 months long for the population to maintain its size from one year to the next. A longer breeding season will result in an increase in population size by the end of the year, but a shorter breeding season will result in a decrease in population size by the end of the year.

The parameter $P'_{12-t}e^{rt}$ is the number of times that the population will increase in one year's time, and it varies with the length of the breeding season (Table 2.3).

The replacement rate for a seasonally or intermittently breeding population whose l_x and m_x schedules are constructed on a weekly or monthly interval is not the reciprocal of the sum of l_x values [Eq. (2.13)], as it is for populations that breed continuously or whose l_x schedule is based on an annual interval. The annual replacement rate, that is, the birth rate during the breeding season that will maintain a population's size from one year to the next, is a function of r and the length of the breeding season. For population A in Table 2.3, which breeds for 5.25 months, the replacement rate is 0.4816, which corresponds to $r'_a = 0$. A longer breeding season will lower the replacement rate, and a shorter breeding season will increase it.

Leslie and Ranson (1940) applied this analysis to a laboratory population of the vole (*Microtus agrestis*), an animal that is active in the field throughout the year. They assumed that l_x schedules during the breeding and nonbreeding seasons were not different. Probably, however, l_x schedules during the nonbreeding season are different from those of the breeding

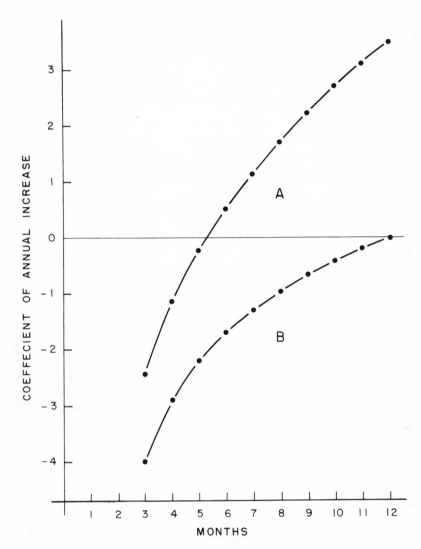

Figure 2.1. The coefficient of annual increase (r'_a) of intermittently breeding populations. Population A with $r = 0.3010$ maintains a constant size from one year to the next if it breeds for 5.25 months. A shorter breeding season results in population decline despite its positive r. Population B with $r = 0.000$ must breed continuously throughout the year or decline.

season, especially for those species that overwinter as diapausing eggs, dormant larvae, or hibernating adults whose lowered metabolism, inactivity, and seclusion in protective sites make them less vulnerable to death from hunger, cold, or predation. Regardless of the difficulties involved in accurately quantifying the population parameters of a particular population, it should be clear that r and R_0 do not provide a measure of population growth for intermittently breeding populations whose l_x and m_x schedules are based on intervals of less than 1 year.

Sampling for l_x and m_x Data

I have chosen to illustrate the mathematics of demography with a hypothetical example in order to avoid dealing with the additional assumptions, sampling errors, and so on, associated with application of the equations to both field and laboratory data, as each population presents unique problems of interpretation. The essential data set, the l_x and m_x schedules, is difficult to establish. Caughley (1966) discussed the assumptions and difficulties involved in obtaining mortality data, whether l_x (survivorship), q_x (the death rate for the interval x, $x + 1$), or d_x (the fraction of a cohort that dies during the interval x, $x + 1$). He showed that mortality data are usually developed from (1) recording the ages of death of a large number of animals born at the same time, giving a kd_x series (where k is a constant that converts the proportion d_x to numbers), (2) recording the number of animals of a cohort still alive at ages x (kl_x series), (3) recording the ages at death of marked individuals not necessarily members of a cohort (kd_x series), (4) recording the ages of death from carcasses of a population assumed to be in steady state (kd_x series), (5) recording the ages of death of a sample killed by a catastrophic event (sometimes gives a kl_x series), and (6) recording the ages of living animals of a steady-state population by trapping or shooting (sometimes gives a kl_x series). The last three methods require the populations to be in steady state. It is virtually impossible, however, as Caughley (1966) pointed out, to determine whether a population is in steady state or has a stable age distribution from a single sample. The same l_x schedule can generate very different mortality rates [compare columns (2) and (11), Table 2.1] depending on whether the population is growing or not. Also, the same table of age-specific mortality can occur with different l_x schedules (Table 2.4), again depending on whether the population is growing or not.

The best estimates of mortality, methods (1), (2), and (3), however, can be misleading in providing information regarding the population's dynamics because although emigration is included (counting disappearances as deaths) immigration is not.

TABLE 2.4

Age-Specific Mortality Rates for Two Populations with Different Growth Rates

Age (x)	Age-specific death rates (q_x)	Survivorship	
		$r = 0.3010$	$r = 0.0$
1	0.3339	1.000	1.000
2	0.3420	0.900	0.666
3	0.3528	0.800	0.438
4	0.3660	0.700	0.283
5	0.3822	0.600	0.179
6	0.4091	0.500	0.110
7	0.4444	0.400	0.065
8	0.5077	0.300	0.036
9	0.6250	0.200	0.018
10	1.0000	0.100	0.007

Estimating m_x values is more difficult than estimating survivorship or mortality because m_x values represent the average reproduction of all the individuals comprising a particular age class. This requires knowing not only the clutch or litter size of the breeding females but the proportion of each age class that is reproducing. The latter is difficult to determine in the field.

Efforts to determine both l_x and m_x data have been confined largely to laboratory populations, such as the human louse (*Pediculus humanus*) (Evans and Smith, 1952), the vole (*Microtus agrestis*) (Leslie and Ranson, 1940), the rice weevil (*Calandra oryzae*) (Birch, 1948), *Daphnia pulex* (Frank *et al.*, 1957), the milkweed bug (*Oncopeltus fasciatus*) (Dingle, 1968), and the flour beetle (*Tribolium castaneum*) (Mertz, 1971a), or to controlled field experiments, such as with aphids (Root and Olson, 1969). Such conditions preclude immigration, emigration, predation, and starvation. Furthermore, l_x and m_x data are calculated during the breeding season. These data, then, can give only a partial explanation of the dynamics of these populations in the field. Data from natural populations generally concern only l_x or d_x data (Deevey, 1947; Caughley, 1966). Field data on clutch size and litter size (Lack, 1954, 1966, 1968) do not provide a measure of m_x, the average reproduction of all members (or females) of a given age of the population. It seems unlikely, then, that an accurate quantitative description of a population can be done. More important than precisely measuring population parameters is establishing *patterns* of mortality and reproduction (Deevey, 1947; Caughley, 1966; Lack, 1968). Lotka's equations allow evaluation of the consequences of different patterns of mortality and reproduction, work begun by Cole (1954).

In this book, we will explore the dynamics of population growth in terms of Lotka's equations. Before we begin, it is useful to identify two relationships as "laws of population dynamics."

The First and Second Laws of Population Dynamics

The First Law of Population Dynamics: Any Population
with Constant Age-Specific Death Rates and
a Constant Recruitment Rate Eventually
Assumes a Steady State

The recruitment rate refers to the number of individuals of age 0 added to the population per annum (or any other convenient time period). Constant age-specific death rates take into account the effects of immigration and emigration. I prefer the term "steady state" to the usual term "stationary state" to describe a nongrowing population because the latter term implies an unchanging rather than a dynamic system.

The law functions in population dynamics as Newton's first law of motion functions in physics and the Hardy–Weinberg law functions in population genetics, that is, it describes an abstract ideal against which the deviations that occur in natural situations can be measured and evaluated. Laws allow us to think about and to analyze what is happening in nature. They do not necessarily tell us what is happening in nature.

This law is a truism, as can be illustrated with a numerical example (Table 2.5). Assuming a recruitment of 100 individuals each year, 10% of which die in each succeeding year, a population, whether of manufactured goods (Boulding, 1955) or animals, will eventually attain a steady-state population of 550, regardless of its initial size.

The first law shows us the conditions for a steady-state population—the age-specific death rates must be constant, and the number of recruits must be constant. The steady-state population size can be increased only by (1) increasing average life expectancy, (2) increasing recruitment, or (3) some combination of (1) and (2).

The relationship between survivorship, the population size, birth rate (or replacement rate), and recruitment of three hypothetical steady-state populations is shown in Table 2.6. The greater the survivorship, the lower the replacement rate, indicating that for any population, whether steady-state or not, an increase in its mortality rate necessarily increases its replacement rate. This relationship will be referred to frequently in later chapters.

Populations with the same recruitment but different survivorship will

TABLE 2.5

Age Structure of Stable and Steady-State Population

Age:	1	2	3	4	5	6	7	8	9		
Survivorship:	1.00	0.90	0.80	0.70	0.60	0.50	0.40	0.30	0.20	0.10	0.00
Time:	0	1	2	3	4	5	6	7	8	9	10
Cohort											
0	100	90	80	70	60	50	40	30	20	10	0
1		100	90	80	70	60	50	40	30	20	10
2			100	90	80	70	60	50	40	30	20
3				100	90	80	70	60	50	40	30
4					100	90	80	70	60	50	40
5						100	90	80	70	60	50
6							100	90	80	70	60
7								100	90	80	70
8									100	90	80
9										100	90
10											100
Totals										550	550

reach different steady-state sizes. Or populations having the same steady-state size but different survivorship will differ in their recruitment (Table 2.7).

Finally, assuming that there is some maximal life expectancy, population growth is limited by recruitment. In order to estimate the prospects of a particular population, for instance determining whether it is growing or declining for management purposes, conventional methods require a fairly accurate census of the population by age class and estimates of age-specific mortality and reproduction. Such information is difficult to obtain, not only for populations of wild animals but for some human populations. Determining the number of eggs laid or babies born may provide a useful and simple estimate of future population growth. If the number of eggs or babies is increasing at 5% a year, the entire population will eventually be growing at 5% a year, assuming constant mortality and stable age distribution.

The Second Law of Population Dynamics: Any Population
with Constant Age-Specific Birth and Death
Rates Will Eventually Establish
a Stable Age Distribution

As already mentioned, this important relationship between survivorship and age structure was developed in a series of papers by Lotka (1907a,b,

TABLE 2.6

Survivorship of Three Populations

Population	l_x											Steady-state population (recruitment = 100)	Average birth rate (per head)
	Age												
	1	2	3	4	5	6	7	8	9	10	11		
A	1.00	0.95	0.90	0.85	0.80	0.75	0.70	0.65	0.60	0.50	0.0	770	1.30
B	1.00	0.90	0.80	0.70	0.60	0.50	0.40	0.30	0.20	0.10	0.0	550	1.82
C	1.00	0.50	0.40	0.35	0.30	0.25	0.20	0.15	0.10	0.05	0.0	330	3.03

TABLE 2.7

Stable Age Distributions of Three Steady-State Populations

Age	l_x	Number	l_x	Number	l_x	Number
1	1.0	130[a]	1.0	182[a]	1.0	303[a]
2	0.95	124	0.9	164	0.5	152
3	0.9	117	0.8	145	0.4	121
4	0.85	110	0.7	127	0.35	106
5	0.8	104	0.6	109	0.3	91
6	0.75	97	0.5	91	0.25	76
7	0.7	91	0.4	73	0.2	61
8	0.65	84	0.3	55	0.15	45
9	0.6	78	0.2	36	0.1	30
10	0.5	65	0.1	18	0.05	15
		1000		1000		1000

[a] Annual recruitment.

1913a,b, 1922; Sharp and Lotka, 1911; Dublin and Lotka, 1925). The frequency of each age class in a stable age distribution is given by Eq. (2.7), $c_x = bl_x e^{-rx}$. The frequency of each age class in a steady-state population is given by Eq. (2.14), $c_x = bl_x$.

Discussions in textbooks of the stable age distribution are incomplete or incorrect. For example, it is frequently stated that growing populations are characterized by a high proportion of young animals. Although this is true for the human populations often used for illustration (but not so identified), it is not necessarily true for other populations. What is true is that growing populations have a higher proportion of young [calculated from Eq. (2.7)] than is the proportion expected in a steady-state population [calculated from Eq. (2.14)] with the same l_x schedule. Stable age distributions of growing, steady-state, and declining populations for each of three l_x schedules (Table 2.6) are shown in Fig. 2.2. Whether a population's age structure gives any indication of the population's increase or decline can be determined only by comparing the actual age structure with the steady-state age structure for a given l_x schedule. A high proportion of young age classes may reflect high mortality of young rather than an increasing population. A declining population with survivorship C in Fig. 2.2 has higher proportions of the youngest age classes than have the growing populations with survivorships A and B. The age structure of a population is more greatly affected by mortality than by reproduction (Coale, 1958).

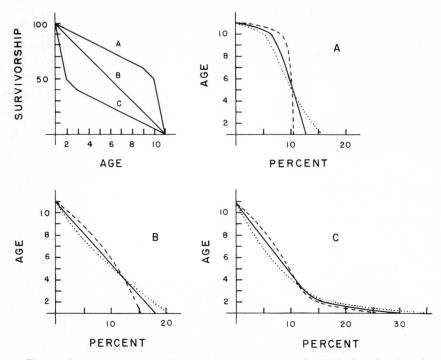

Figure 2.2.   Stable age distributions of nine populations. Graphs A, B, and C show the stable age distributions for growing ($\cdots\cdots$), steady-state (———), and declining (----) populations with survivorship schedules A, B, and C, respectively. Survivorship is shown in the upper left. As can be seen, mortality has a greater effect on the stable age distribution than does fertility. (From Coale, 1958.)

THE PROJECTION MATRIX

The use of the exponential, logistic, and Lotka's equations assumes a stable age distribution. For a population without a stable age distribution, Leslie (1945, 1948) introduced matrix methods for projecting population growth. The data needed are (1) the number of individuals alive in each age group at time t, (2) the number of progeny born during the period t to $t + 1$ that will be alive at time $t + 1$, and (3) the probability that an individual alive at time t will be alive at time $t + 1$. The matrix method is useful for projecting population growth, especially for populations lacking stable age distributions. This method will not be used in the development of the theories in later sections, so it is not discussed further here.

THE EFFECT OF AGE DISTRIBUTION

The underlying assumption for projecting future population growth with either the exponential or logistic equation is that the population has a stable age distribution. Nevertheless, ecologists attempt to apply the logistic to both natural and laboratory populations without first establishing the age structure of the populations. The search for explanations of population fluctuations has focused on time lags (Hutchinson, 1948; Andrewartha and Birch, 1954; Wangersky and Cunningham, 1958; May, 1976a) rather than on age structure.

Inasmuch as the theoretical development in the next few chapters assumes stable age distributions, it is worthwhile to consider the effect on population growth of a nonstable age distribution. Assume steady-state populations with survivorship schedule B (Table 2.6). These populations have constant age-specific birth and death rates, constant sizes, stable age distributions, $r = 0.0$, and $R_0 = 1$. Assume further a one-time change in recruitment. This one-time change results in *persisting* fluctuations (Fig. 2.3), even though l_x, m_x, and, therefore, r remain constant. The alternating periods of positive and negative growth dampen out as the age distribution approaches stability. After perturbation, the age distribution returns to its initial stable age distribution, but note that the size of the populations does not return to its initial steady-state size.

As shown by Lotka, the stable age distribution tends to perpetuate itself. After a perturbation, the age distribution of a population will return to its

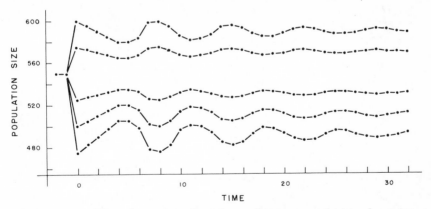

Figure 2.3. Effect of a one-time change in recruitment on growth. Annual recruitment of 100 individuals maintains a steady-state of 550 individuals (Table 2.6). In these populations there is a one-time annual recruitment of 150, 125, 75, 50, and 25 individuals, from top to bottom. In subsequent years recruitment is 100.

original stable condition, if age-specific birth and death rates remain constant. However, any perturbation to a steady-state population will also result in a *change* in the population's steady-state size, if l_x and m_x schedules are perturbed once. These populations cannot be called equilibrium systems. Nor, however, can the stable age distribution be considered an equilibrium system even though it tends to perpetuate itself, because there is no negative feedback loop responsible for bringing about the stable age distribution after a perturbation. The stable age distribution is an inevitable, mathematical consequence of the conditions stated in the first and second laws of population dynamics.

The theoretical development in the next chapter concerns how a population shifts from a period of growth $(r > 0)$ to a period of steady state $(r = 0)$ and how that shift occurs at one density instead of another. Solutions to these questions need not consider perturbations to the age distributions of populations because these perturbations only introduce additional complexity that does not affect the theory but does affect the testing of theory. If theory assumes a stable age distribution and predicts a steady state when field or laboratory data show fluctuations, we do not want to reject theory if the fluctuations are owing to nonstable age distributions. Alternatively, if theory assumes a stable age distribution and predicts fluctuations when field or laboratory data show fluctuations, we do not want to accept theory if the latter fluctuations are owing to nonstable age distributions. Unfortunately, the age distributions of few populations are known, making the testing of theory not unambiguous.

3

LIMITATIONS TO POPULATION GROWTH

A population's size increases (phase A in Fig. 1.2) when recruitment or life expectancy increases or both increase. This occurs when immigration exceeds emigration and when the birth rate exceeds the replacement rate. Thus, the shift from the growth phase to the steady-state phase (B in Fig. 1.2) involves a decline in immigration, an increase in emigration, a decreased birth rate, an increased death rate, or some combination of these. Because we are interested in the dynamics of a population within a defined space, emigration from the area can be counted as deaths, and immigration can discount age-specific deaths depending on the age of the immigrants. We can, then, describe the net effect of deaths, emigration, and immigration on a population within a defined area by an l_x schedule. This survivorship can be translated into a single index number, the replacement rate [Eq. (2.13)]. Whether or not a population is growing is determined by comparing the population's birth rate with its replacement rate. At steady state the birth and replacement rates are equal.

Lack (1954, 1966) and his many followers have maintained that natural selection maximizes individual reproduction and that, therefore, a steady state is achieved by density-dependent factors acting on death rates. Others (Skutch, 1949, 1967; Wynne-Edwards, 1962) have suggested the opposite, that reproductive rates change in response to environmentally induced mortality. Emigration (i.e., dispersal) is sometimes considered a means of preventing overpopulation (Lidicker, 1962; Wynne-Edwards, 1962).

The alternatives to the density-dependent regulation theories, the theories of Thompson (1939, 1956), Andrewartha and Birch (1954), Schwerdtfeger (1958), and in part Milne (1962), suggest that population growth is limited by the availability of resources or of time in which to reproduce and that fluctuations are often the result of density-independent changes in the environment.

Neither the regulation theories nor the limitation theories have been developed in terms of the interrelated effects of one population parameter on the others. As indicated earlier, for example, a change in age-specific death rates affects the l_x schedule, stable age distribution, and replacement rate, and it can affect the birth rate as well. The following sections of this chapter discuss the cause and effect relationships of these parameters with respect to space-limited, time-limited, food-limited, and predator-limited populations.

SPACE AS A LIMITING FACTOR

A factor limiting annual recruitment in some populations is territorial behavior. Because there seems to be a limit to which territories can be compressed (Huxley, 1934), only so many individuals can establish territories within the available habitat and successfully breed. Territorial aggression, then, limits recruitment by limiting the number of individuals that can establish territories and by limiting the number of births.

A numerical example will show how a sigmoid-shaped growth curve can be generated in a territorial population even if no change occurs in either age-specific death rates (i.e., survivorship is constant at all densities) or individual fecundity. Assume (1) the survivorship

Age:	1	2	3	4	5
l_x:	1.0	0.75	0.50	0.25	0.0

(2) that each breeding female produces 2.0 young per year at any density, and (3) that territorial behavior allows a maximum of 25 pairs to breed within the study area. The replacement rate for a population with this survivorship is 0.4 [from Eq. (2.13)], which means that each female must produce on average 0.8 young per year. If the females are producing 2.0 young per year, the population will grow. However, the breeding population can grow only until 25 pairs occupy territories (Fig. 3.1). The total population continues to grow (Fig. 3.1), but because the maximum number of young that 25 breeding females can produce is 50, the population's birth rate declines (Fig. 3.2). The population stabilizes at 125 individuals, con-

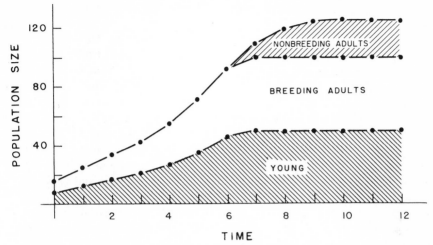

Figure 3.1. Growth of a population limited by territorial behavior. The minimal territory size allows a maximum of 25 males to establish territories within the area under study. The survivorship schedule of this population, regardless of density, is l_1, 1.00; l_2, 0.75; l_3, 0.50; l_4, 0.25; l_5, 0.00. The replacement rate, therefore, is 0.40. The reproductive ages are 2, 3, and 4, and each breeding female produces 2.00 young at each reproductive season, regardless of density. The species is monogamous. Nevertheless, the population shows sigmoid growth. This comes about because territorial behavior limits recruitment.

sisting of 50 immatures, 25 pairs of breeding adults, and 25 nonbreeding adults (often called "floaters").

The initial changes in the population's crude birth and death rates (Fig. 3.2) represent the effects of the shift from the nonstable age distribution of the founder population toward stability (second law of population dynamics). The later changes in birth and death rates are the result of the shift from the stable age distribution of the growing population to the steady-state age distribution. Even though the changes in birth and death rates are "density-dependent" in the sense that the birth rate declines and death rate increases (slightly) within the upper range of increasing densities, these changes cannot in any way be construed to be the cause of the population's changing from a growth phase to a steady-state phase. Instead, the density-related changes in birth and death rates are consequences of the changing age distribution imposed on the population by the limited number of territories. The population size, then, is determined by the number of pairs that are able to establish territories and the number of young each pair raises.

The importance of this model is that it provides a reasonable explanation for the determination of the numbers of a steady-state population without resorting to density-dependent factors, apparently the first model to do so (Richards and Southwood, 1968). Certainly, many authors have suggested

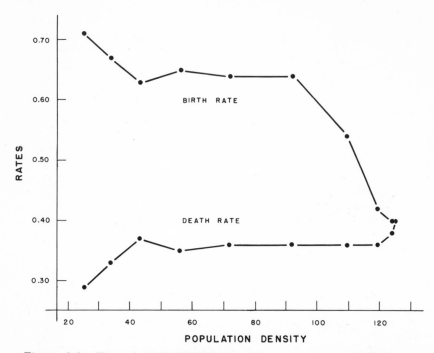

Figure 3.2. The crude birth and death rates of the territorial population shown in Fig. 3.1. Despite unchanging survivorship and unchanging fecundity of the breeding females, the population's crude birth and death rates vary. The changes occur because the population's age distribution is changing in time, at first because of the unstable age distribution of the founder population and later (above 90) because territorial behavior limits the number of breeders, causing the nonbreeding population to increase, the birth rates to decrease, and the age distribution to change once again.

that territoriality sets the upper limit to population density (Huxley, 1934; Kluyver and Tinbergen, 1953; Tinbergen, 1957; Tompa, 1964a; Wynne-Edwards, 1962; Watson, 1967; Watson and Jenkins, 1968; Watson and Moss, 1970; Harris, 1970a; Fretwell and Calver, 1970; Krebs, 1971; Klomp, 1972; S. M. Smith, 1978), but the existence of density-dependent *effects* on territory size, clutch size, and reproductive success have served to confuse the understanding of the dynamics of populations of territorial species.

Lack (1954) argued that, if territories do in fact limit populations, then territory size should be nearly constant within a species, the minimal size of territories must frequently be reached, and the minimal size must correspond to the minimal amount of food that is needed by a pair to raise their young. There is no evidence, as Lack pointed out, for any of these conditions, and, therefore, Lack concluded that territoriality does not limit populations. Unfortunately, Lack did not give his reasons for setting these conditions, which seem to others too restrictive for the problem being discussed (Hinde, 1956; Tinbergen, 1957; Watson and Moss, 1970). For territoriality to *limit* population size, the only requirement is for territorial behavior to prevent some intruding individuals from establishing territories in the area that includes the population of interest.

That territoriality may act as a limiting factor does not mean that all territorial species are always at the limit. A founder population of a territorial species will be small relative to the size that the habitat can support and the minimal territory size will permit. At low density, the size of the territories may be large. As the population grows and becomes denser, the territories become smaller until the minimum size is reached (Fig. 3.3), after which time recruitment is constant. At no time does territoriality act as a density-dependent factor, even though territory size and the population's crude birth rate vary with density. While territory size is decreasing,

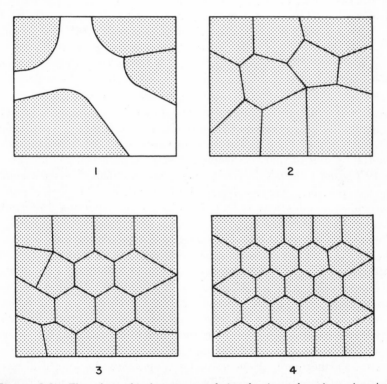

Figure 3.3. The relationship between population density and territory size. As the population size increases, the size of the individual territory declines. At low densities, the entire habitat may or may not be completely occupied by the species, and territories need not be of uniform size, even at the maximum density.

age-specific birth rates and death rates remain constant, and, while birth
and death rates are changing, the territory size remains constant. There is
no correlation between changes in birth and death rates and changes in
territory size.

The literature on territoriality records many observations indicating that
territorial behavior was probably responsible for forcing some individuals
to breed in less suitable habitats or for preventing some from breeding at
all (Hinde, 1956; Brown, 1969a). Replacement of territorial birds that have
been shot, accidentally killed, or otherwise removed has been documented
(Stewart and Aldrich, 1951; Hensley and Cope, 1951; Orians, 1961;
Holmes, 1966; Watson and Jenkins, 1968; Harris, 1970a; Krebs, 1971;
Murray and Gill, 1976; Morton, 1977; Thompson, 1977). By means of
color-marking, nonbreeding, nonterritorial individuals ("floaters") have
been identified in some populations (Kendeigh, 1941; Carrick, 1963; Delius, 1965; Harris, 1970a; S. M. Smith, 1978). In a few species, some
individuals were found to establish territories in habitats where the probability of successful reproduction was low compared with that of other
individuals (Fretwell and Calver, 1970; Zimmerman, 1971; Krebs, 1971).

Brown (1969a) raised the important point that territorial behavior may
cause a surplus of males but not necessarily a surplus of females. If all
females breed, the territorial behavior of males is largely irrelevant with
respect to determining the size of the population. In only a few studies has
an apparent surplus of females been demonstrated (Carrick, 1963; Delius,
1965; S. M. Smith, 1978), but in my work with color-marked populations of
territorial species of sparrows (Murray, 1969), wood warblers (Murray and
Gill, 1976), and a thrush (B. G. Murray, Jr., unpublished observations), I
found that even *breeding* females are difficult to detect. The lack of data on
nonbreeding females may reflect the difficulty of detecting them rather
than their nonexistence.

There is no reason to believe that a territorial population must always be
at its limit or that any population with "floaters" is at its maximum limit.
Territory size can vary seasonally and spatially (Tinbergen, 1957; Watson
and Moss, 1970; Klomp, 1972) and is "a function of the *interactions*
between the quality of the environment, the condition and number of
competing individuals, and the behavior of the species" (Brown, 1969a; see
also Klomp, 1972). As minimal territory size varies with repect to these
interactions, recruitment will fluctuate and result in a fluctuating population size. Furthermore, the nonbreeders may leave the area if dispersal
leads to increasing their probability of successful reproduction elsewhere
(Murray, 1967).

A striking example is Tompa's (1964a,b, 1971) study of a marked Song
Sparrow (*Melospiza melodia*) population on Mandarte Island in British

Columbia. The number of *breeding* territorial pairs over a 4-year period was 47, 47, 44 and then 69. Following the first three seasons, Tompa (1964a) suggested that territoriality was the main factor determining the Song Sparrow population size. On the basis of the fourth season's results (Tompa, 1964b), Lack (1966) argued that territorial behavior was ineffective in limiting population size. In a subsequent analysis of unpublished data, Tompa (1971) showed that the situation on Mandarte Island was more complex than he had reported earlier and more complex than Lack had expected.

During the peak of Song Sparrow territorial activity in the third breeding season, a severe snowstorm occurred. Prior to the storm, there was a minimum of 44 established territorial pairs. In addition, there were about 54 young males, one-third of which were territorially active. All young females were mated to the territorial adult males. The storm caused high mortality, with losses amounting to 34% of territorial adult males, 36% of adult females, 42% of young but mated females, and 72% of the territorially active young males. All territorially inactive young males survived. This mortality resulted in the breaking up of well-established neighbor relationships, thus allowing the subordinate young males to establish territories in areas from which they had previously been excluded. All the marked young males eventually established territories, but only 44 of the 61 territorial males were able to acquire mates, immigration of females just making up for the losses. Another 8 males remained as "floaters." Further immigration in the autumn, a period of dispersal and spacing activity in Song Sparrows, made up for the summer deaths and for the low proportion of females, such that at the beginning of the following breeding season, there were 69 breeding pairs. Thus, as Tompa (1971) concludes, contrary to Lack, territorial behavior can and does limit a population's size. The differences in the upper limit at different times result from the conditions under which territories are being established.

The model presented here for territorial species is applicable to any space-limited population. Mammals do not often establish mutually exclusive spaces called territories. Nevertheless, aggression among the individuals within a space prevents some individuals from breeding within that space (Sadleir, 1965; Healey, 1967). In some hole-nesting species, such as bees (Andrewartha and Birch, 1954), aggression is restricted to near the nest site, but it has the same effect on population dynamics as territoriality.

In this section, I have shown how territorial behavior can limit population size and growth without recourse to density-dependent factors. The model generates a steady-state population consisting of (1) young, (2) breeding adults, and (3) nonbreeding adults ("floaters"). There is an abundance of data that are consistent with the predictions of the model (see papers cited above), and I therefore consider the model to have strong support. Most previous discussions of the role of territorial behavior in limiting populations have suffered from their authors' preoccupation with creating density-dependent explanations for the determination of the population's numbers (Lack, 1954, 1966; Krebs, 1970, 1971; Klomp, 1972), with the function(s) of territoriality (Brown, 1964, 1969a), or with group selection theories to explain the evolution of territorial behavior as a means of preventing overpopulation (Wynne-Edwards, 1962). As shown here, however, steady-state, territorial populations can be explained without reference to density-dependent factors, the function(s) of territorial behavior, or group selection.

Confusion has also been caused by failure to define the population as a group of animals living in a particular area. Brown (1969a), Krebs (1971), and Thompson (1977) suggest that, although territorial individuals exclude intruders from breeding within their territories, territorial behavior does not limit or regulate the population's size if the intruders are able to breed elsewhere. This is another example of the area effect, discussed in Chapter 1.

Brown (1969a,b) recognized three critical levels of population density with respect to territorial behavior. At level 1 only optimal habitat is occupied (Fig. 3.4). When optimal habitat is full, level 2 is reached, and suboptimal habitat begins to be occupied. Eventually, the suboptimal habitat is full, and the individuals that are unsuccessful in establishing territories in either optimal or suboptimal habitat remain as "floaters" or disperse (level 3). According to this conception, territorial behavior limits the population living in optimal habitat when individuals are forced to live

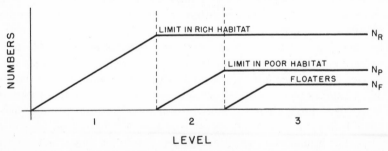

Figure 3.4. Hypothetical effects of territoriality on breeding densities in rich and poor habitats. N_R is the number of breeders in rich habitat; N_P is the number of breeders in poor habitat; N_F is the number of floaters. The total population is the sum of N_R, N_P, and N_F. At level 1, birds breed only in rich habitat until the limit set by minimal territory size is reached. Then, at level 2, birds begin breeding in poor habitat until the limit there is reached. After all territories are occupied, birds can remain as floaters (level 3). (Modified from Brown, 1969.)

in suboptimal habitat, and territorial behavior limits the population living in optimal and suboptimal habitats when some individuals become "floaters." Whether a population is considered limited by territorial behavior is clearly dependent on area size. If the study area is located in optimal habitat, we might observe that some individuals are prevented from breeding by territorial behavior. We would not be incorrect to conclude that territorial behavior is limiting the population we are studying, even if we see the losers breeding elsewhere. If we expand our study area to include suboptimal habitat as well as the optimal habitat, we might observe that the excess from the optimal habitat breeds in the suboptimal habitat, that the population in suboptimal habitat fluctuates in size, and that there are no "floaters." We would not be incorrect to conclude that territorial behavior was not limiting the population in the larger study area. In this case territorial behavior limits the density of that portion of the population living in optimal habitat (see also Klomp, 1972). Some other factor must be limiting the total population.

FOOD AND PREDATION AS LIMITING FACTORS

The ultimate factor (not in the evolutionary sense) limiting population growth and size is the amount of food available for consumption. No population can exceed the size allowed by its energy supply, a simple consequence of the second law of thermodynamics, except temporarily during times of famine. Famines, of course, occur when food supplies become reduced, either by the population's own growth or by adverse environmental changes, and result in the population's decline.

Andrewartha and Birch (1954), Andrewartha (1958), and Andrewartha and Browning (1961) have shown that the interaction between a population and its food supply is more complex than usually treated. First, food is a consumable resource, whereas other factors that affect survivorship, fecundity, and growth are not. Food resources need to be replaced as they are eaten. If they are not, the population will certainly decline. Second, producer, decomposer, and scavenger populations have different effects on their food supply than do herbivores and carnivores. Producers, decomposers, and scavengers feed on inorganic and nonliving organic matter, respectively, Thus, additions to the supply of food for producers, decomposers, and scavengers are independent of the consumer's population density. Herbivores and carnivores, however, feed on living populations, and their consumption will affect the rate of growth of their food supply by changing survivorship and, perhaps, the fecundity of their food populations. When consumers prevent their food populations from growing to the limit set by their own food resources, the consumed population can be considered predator-limited or herbivore-limited, depending on whether the consumed population is animal or plant.

Figure 3.5 shows the relationship between the birth and replacement rates of a food-limited consumer population and the population's density. The assumptions are that (1) there is a maximal birth rate at each age (m_x schedule) and a minimal death rate at each age (l_x schedule), which is reflected in the replacement rate ($b_r = 1/\Sigma\, l_x$), set by the inherent characteristics of the individuals and by the prevailing conditions of other environmental factors (e.g., temperature, predation); (2) below a lower critical density (LCD), birth rate declines because of undercrowding effects (Allee *et al.*, 1949); (3) above an upper critical density (UCD), declining per capita food consumption causes both birth rate and survivorship to decline; and (4) between the LCD and UCD, birth and replacement rates are constant. The population's growth is slow below the LCD, maximal between the LCD and UCD, and above the UCD it slows again, but

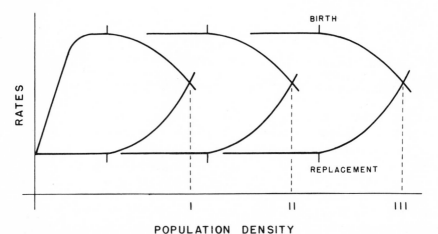

POPULATION   DENSITY

Figure 3.5. Birth and replacement rates of food-limited populations. Shown are birth and replacement rates for three food-limited populations of the same species, each located in an area with a different amount of available food. At low and moderate densities, per capita consumption is maximal; therefore, the birth rate is maximal (except at the lowest densities, when the rarity of individuals reduces the chances of meeting mates and reproducing successfully), and the mortality and replacement rates are minimal. As the population increases, at some point (the upper critical density, indicated by the tick marks on the curves) per capita consumption begins to decline, causing an increase in the replacement rate and a decrease in the birth rate. The UCD occurs at different densities, depending on the food supply. Continued population growth further reduces the birth rate and increases the replacement rate. Finally, birth and replacement rates are equal, and the population stops growing at different densities (I, II, III), depending on the food supply. (This is the basic model, which is further developed and modified in the text and in later figures.)

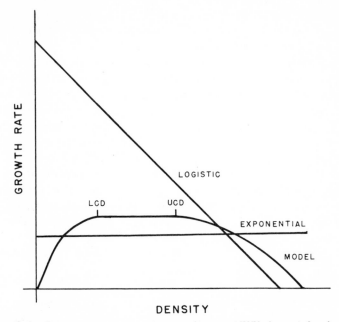

Figure 3.6. Instantaneous per capita growth rates ($dN/N\ dt = r$) for three model populations—exponential growth, logistic growth, and the model presented in Fig. 3.5. In the last, LCD refers to lower critical density and UCD to upper critical density.

numbers continue to increase. As intraspecific competition for food intensifies, a point is reached when the population's birth rate falls below the replacement rate, and the population stops growing. In this model, the population's growth rate (r) is not a linear function of density (Fig. 3.6).

The UCD is different for each population. It depends on the amount of food available for consumption. Populations living in food-rich areas will have larger UCDs and will attain larger sizes than populations living in food-poor areas (Fig. 3.5). It is not correct to think of intraspecific competition as a density effect on survivorship and reproduction. It is instead a crowding effect, and crowding occurs at different densities.

This, then, is the basic model for food-limited populations. Although in this model (Fig. 3.5) I have shown mortality as minimal at the lowest densities, it is possible that mortality may increase at the lowest densities because, for instance, individuals in small groups may be less able to protect themselves from predators or climate as well as individuals in larger groups.

Pielou (1977) presented a relationship between r and density similar to that shown in Fig. 3.6 as a "modification" of logistic growth. This relationship, however, is not a modification but is instead a different model with considerably different biological properties. In the logistic, r ($= dN/N\ dt$) declines linearly with increasing density (Fig. 3.6). The logistic implies a negative feedback between a population's density and its per capita growth rate. No such negative feedback is implied by the model described above (Fig. 3.5). Below the LCD, growth rates are low because of "undercrowding" effects, and, above the UCD, growth rates are low because of "overcrowding" effects (intraspecific competition for food). Between the LCD and UCD, the growth rate is maximal and remains constant over a range of densities, the extent of the range depending on the food supply. This model is intuitively more appealing than either exponential or logistic growth models and more importantly offers a testable alternative to the logistic model of population growth.

Pielou (1977) showed that the increase in numbers of the model population is sigmoid, as in the logistic growth model. The two sigmoid curves, however, are distinguishable, in theory if not always in practice. In the model population, the growth rate r is constant through the intermediate densities (Fig. 3.6), whereas, in logistic growth, the growth rate is different at all densities (Fig. 3.6). The two models generate different patterns of growth in numbers of food-limited populations exposed to five different levels of food (Fig. 3.7A and 3.7B). In my model, numbers grow at the same rate at all densities until the UCD of each population is reached (Fig. 3.7A), whereas, in logistic growth, growth rates are different between populations at the same density (Fig. 3.7B). The actual growth of six beetle (*Tribolium confusum*) populations given different levels of food supply (Fig. 3.7C) and nine guppy (*Lebistes reticulatus*) populations given three different levels of food supply (Fig. 3.7D) does not clearly fit either growth model. Both models assume a stable age distribution, and thus the growth of natural and experimental populations, which may often lack a stable age distribution, may diverge from predicted values, making testing of either model difficult.

Producer, decomposer, and scavenger populations consume nonliving food, and the size of these populations will depend on the food supply. At some point, per capita food consumption declines, leading to increasing mortality and replacement rates and decreasing birth rates (Fig. 3.5). As numbers increase, per capita food consumption continues to decline until the population's birth rate falls below the replacement rate, and the population stops growing. However, if producers, decomposers, and scavengers consume their food faster than it is naturally replaced, the birth and replacement rate curves in Fig. 3.5 will shift to the left, and the population will decline. An influx of food will shift the curves to the right, and the population will increase. The supply of minerals to producers and nonliving organic matter to decomposers and scavengers is independent of the

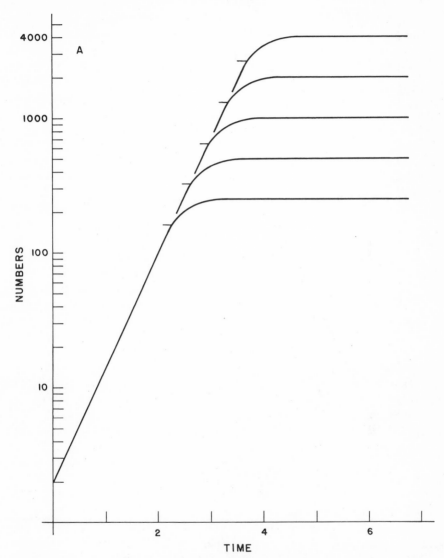

Figure 3.7. Comparison of growth curves. (A) Growth of five hypothetical populations in response to different amounts of food, according to the model proposed in this book (Figs. 3.5, 3.6). (B) Growth of five hypothetical populations in response to different amounts of food, according to the logistic model. (C) Growth of six *Tribolium confusum* populations in response to different amounts of food, from 4 to 128 gm (data from Chapman, 1928). (D) Growth of nine guppy populations in response to three different amounts of food, provided in the ratio 1 : 2 : 3. The data on the three populations at each food level have been averaged (data from Silliman, 1968).

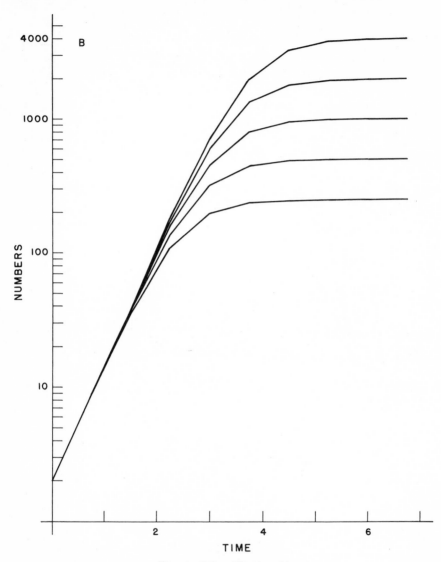

Figure 3.7. (*Continued.*)

consumer's population size (Andrewartha and Birch, 1954; Andrewartha, 1958; Andrewartha and Browning, 1961), and it is likely to vary in time. As a result, food-limited producer, decomposer, and scavenger populations are likely to fluctuate in size in response to variations in the supply of food.

Herbivore and carnivore populations consume living plants and animals, respectively. Herbivores and carnivores interact with their food popula-

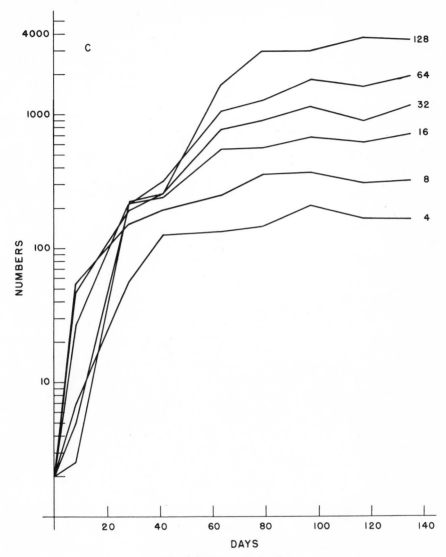

Figure 3.7. (*Continued.*)

tions in similar ways, there being no dynamic difference between herbivory and predation (Andrewartha and Birch, 1954; Andrewartha, 1958; Andrewartha and Browning, 1961). We can, then, illustrate the dynamics of these interactions by considering predator and prey populations.

Both the predator and prey populations are living, and thus each population will have its own birth and replacement rate curves (Fig. 3.8). The

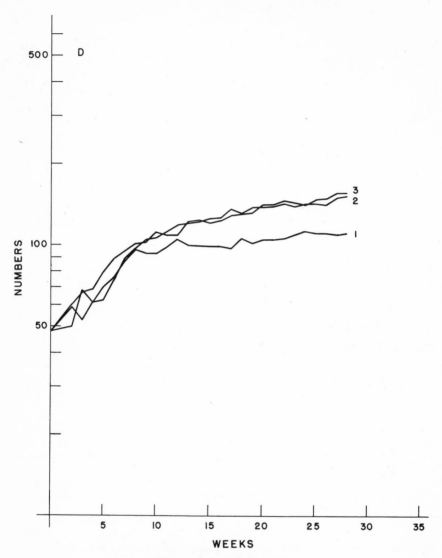

Figure 3.7. (*Continued.*)

predator population, we will assume, is food-limited, that is, survivorship and fecundity decline with increasing intraspecific competition for prey (Fig. 3.8a). Consumption of the prey by the predator, however, imposes an additional mortality on the prey population, causing the replacement rate to increase (i.e., shift to the left in Fig. 3.8b). This reduces the maximal potential size of the prey population, which in turn shifts the birth and

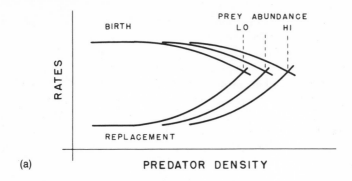

(a)

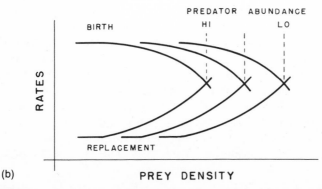

(b)

Figure 3.8. Predator–prey interactions. (a) The size of a food-limited predator population is determined by the available food supply, being greater at high prey densities. (b) The size of a predator-limited prey population, however, is determined, in part, by the size of the predator population, being smaller at high predator density. As a result, an increasing prey population leads to an increasing predator population. This lowers the prey populations, and the predator population declines, allowing the prey population to increase again. This interaction leads to more or less cyclic fluctuations in the sizes of both populations.

replacement rate curves of the predator to the left because intraspecific competition of predators for prey occurs at a lower predator density (Fig. 3.8a). The predator population declines in size, reducing mortality on the prey population, whose replacement rate curve shifts to the right, and the prey population grows. These alternating shifts of the curves establish persisting fluctuations in the sizes of both the predator and prey populations, fluctuations that have been observed in the field (Elton, 1942; Keith, 1963) and the laboratory (Utida, 1957; Huffaker, 1958b) (Fig. 1.6).

The magnitude of the fluctuations will depend on the intensity of the herbivore or carnivore population's consumption and its food population. Some consumer populations may grow rapidly and deplete their food supply more severely than slower growing populations do, and thus their populations will fluctuate with greater amplitude. Both the magnitude and frequency of the fluctuations in size of food-limited herbivore and carnivore populations are likely to be more regular (giving the appearance of being cyclic) than those of producer, decomposer, and scavenger populations inasmuch as the former populations influence each other's numbers, whereas the latter populations' food supply is independent of the consumers' activity.

The argument presented here refers only to food-limited herbivore and carnivore populations. Herbivore and carnivore populations can be limited by territoriality (discussed in the last section) or by time in which to reproduce (discussed in the next section), the predator populations never growing to the size necessary to depress the numbers of their food populations. In such cases, the prey population can grow to the limits set by its own resources (Fig. 3.9). For example, Buckner and Turnock (1965) studied the effects of different densities of the larch sawfly (*Pristiphora erichsonii*) on avian populations. Forty-three of the 54 species of birds sampled were predators on the sawfly. Census data on 34 species showed that most avian populations increased with greater densities of sawflies (Fig. 3.10). Nevertheless, the percentage of sawflies taken by predators declined with increasing density (Table 3.1). The greatest increases oc-

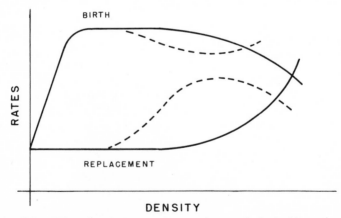

Figure 3.9. Effect of nonlimiting predator on the growth of a food-limited prey population. The birth and replacement rates of prey in the absence of predators are shown by the solid line. The effects of the predators' numerical and functional responses on the prey's replacement and birth rates are shown by the dashed lines. The prey's growth rate is reduced by the predation, but the prey population reaches the same maximal population that it would have reached in the absence of predators (compare with Fig. 3.11).

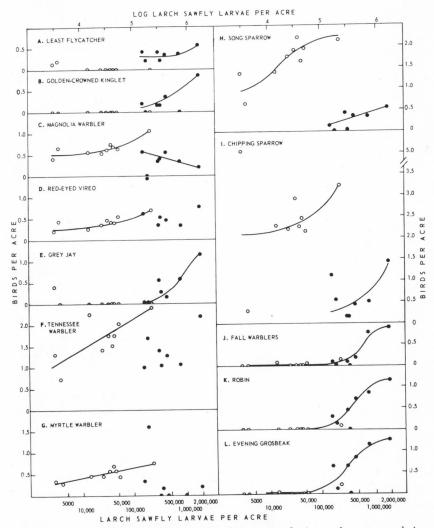

Figure 3.10. Some examples of numerical responses of avian predators to populations of larch sawfly larvae. (From Buckner and Turnock, 1965.)

curred in nonresident, migratory species, but territoriality limited the populations of resident species. Thus, birds did not limit the further growth of sawfly populations.

Birds also were unable to limit the further growth of spruce budworm (*Choristoneura fumiferana*) populations, although some avian populations increased with increasing budworm numbers (Morris *et al.*, 1958).

TABLE 3.1

Predation on Sawfly Populations[a]

	Sawfly population	
	1	2
Prey population/acre		
Larvae	2,138,700	40,000
Adults	205,400	4,700
Potential predation[b]		
Larvae	10,665 (0.5%)	2,344 (5.9%)
Adults	10,601 (5.6%)	3,049 (64.9%)
Predator population/acre	23.5	12.6
Number predator species	28	10

[a] From Buckner and Turnock (1965).

[b] Calculated by considering the metabolic requirements, caloric equivalents of larch sawfly, utilization of sawfly material, functional responses, and predator populations.

The above discussion and model (Fig. 3.8) further assumed a single predator species feeding on a single prey species. This is undoubtedly rare in nature, the closest approach to such a condition occurring in communities of low species richness. For example, at Point Barrow, Alaska, the major predators are the Pomarine Jaeger (*Stercorarius pomarinus*), Snowy Owl (*Nyctea scandiaca*), Parasitic Jaeger (*Stercorarius parasiticus*), Short-eared Owl (*Asio flammeus*), Glaucous Gull (*Larus hyperboreus*), least weasel (*Mustela rixosa*), and arctic fox (*Alopex lagopus*), and all feed mainly on the brown lemming (*Lemmus trimucronatus*). The numbers of both predator and prey species fluctuate greatly. When prey populations are dense the predator populations are also dense (Pitelka *et al.*, 1955; Maher, 1970, 1974), but whether the subsequent decline in prey numbers is a result of predation or other factors is disputed (Thompson, 1955; Pitelka *et al.*, 1955; Pitelka, 1958; Chitty, 1960; Pearson, 1966). Some work indicates that avian predators seem to truncate peak prey numbers, whereas the prey population low is brought about by mammalian predators, in particular, weasels (Maher, 1967, 1970).

In communities of high species richness, predators can shift from prey species to prey species or prey on a greater variety of species, and therefore predator numbers need not fluctuate as greatly as they do in simpler communities. Predators, however, do not prey on the entire community. "Optimal foraging theory" suggests that predators rank prey types according to the ratio between food value (measured in calories or nutritional content) and handling time, that prey are added to the predator's diet by rank order, and that prey of low rank order are excluded from the diet regardless of its abundance (Emlen, 1966; MacArthur and Pianka, 1966; Pulliam, 1974; Krebs and Cowie, 1976; Pyke *et al.*, 1977). Recent experimental and field research tends to support these expectations in the Redshank (*Tringa totanus*) (Goss-Custard, 1977a,b), Spotted Flycatcher (*Muscicapa striata*) (Davies, 1977), and Great Tit (Krebs *et al.*, 1977). These studies were designed to test the "optimal foraging theory" and have not been concerned with the effects of differential prey selection on population numbers.

Predator–prey cycles in species-rich communities have not been studied in detail or, perhaps, at all (Lack, 1954). There is an unpublished report that in the foothills and mountains of northern Alaska, where several species of microtines occur together, the fluctuations in their populations are independent of each other and have no regular periodicities (Pitelka, cited in Maher, 1974).

These considerations suggest that the relationship between predator and prey is complicated by (1) whether the predator population is limited by territorial behavior, by the lengths of the reproductive and nonreproductive seasons (see next section), or by its prey, (2) the species richness of its community, and (3) the ratio of food value to handling time of its potential prey species. We should not expect a single dynamic relationship between predator and prey populations.

Solomon (1949) suggested that predators may change their rate of predation as prey densities change (the functional response) or increase in numbers as prey numbers increase (the numerical response) or both. Holling (1959) showed that in three mammal species, the typical functional response was a sigmoid-shaped increase in consumption with increasing prey density, but the numerical response was more variable. The total response was dome-shaped (Fig. 3.11). Holling was concerned with the predator's role in regulating prey population sizes and developed a simple model by assuming that a fixed percentage of predation could balance the birth rate of the prey, and he drew a line representing this percentage through the total response curve (line x in Fig. 3.11). This assumption and line are completely arbitrary. The percentage of predation that balances the birth rate may be greater than can be achieved by the predators (line x' in Fig. 3.11). In the latter case, predation cannot limit the further growth of the prey population.

Furthermore, Holling's model has the peculiar property that if the prey species manages to run the gauntlet (between A and B in Fig. 3.11), say by a surge of reproduction that is more rapid than the predator's response, then the predator can no longer "regulate" the prey population's size. The prey population continues to increase to the size set by its available food supply. When prey populations do continue to grow despite the functional and numerical responses by its predators, it seems more probable that the

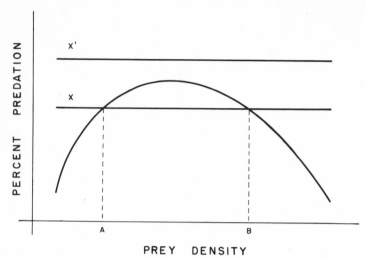

PREY   DENSITY

Figure 3.11. Predator response to prey density. The total predator response to prey density is shown by the curved solid line. The line x indicates the percent predation that will limit further growth of the prey population. The prey population will grow to density A. If the prey population should reach a density between A and B, predation will reduce the prey population to A. If the prey population should exceed density B because of a surge of reproduction, the predator will not be able to limit further growth of the prey population. I have added line x' to show a more probable situation when predators are ineffective in limiting populations (but see text and Fig. 3.9). (Modified after Holling, 1959.)

predators' total response is not sufficient to limit the prey's further growth, that is, the total response does not reach x' in terms of Holling's model (Fig. 3.11). I believe a better representation of the predator–prey interaction is shown in the models presented here. The predator's total response does limit the prey population's size in one case (Fig. 3.8) but not in another (Fig. 3.9).

Several authors have classified the types of predation, from the 2 of Errington (1946) to the 12 of Holling (1959). These attempts at distinguishing different kinds of predator–prey relationships are important for understanding the nature of predation, but so far as population dynamics is concerned, there is only one relationship—the individuals of one species eat the individuals of another, resulting in higher mortality and greater replacement rates for the prey population. The dynamics of predator–prey relationships is examined in greater detail in Chapter 5.

Food as a Regulating Mechanism

It is difficult to see how food can act as a regulating mechanism, in the sense that regulating mechanisms maintain a relatively constant popula-

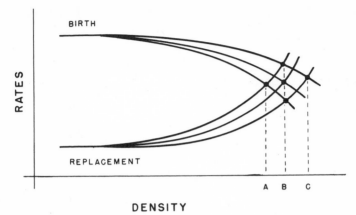

Figure 3.12. Conditions for steady-state populations. For a population (B shown here) to remain constant in size, an increase in fecundity (birth rate curve shifts to right) must be offset by an increase in mortality (replacement rate curve shifts to left). Similarly, a decrease in fecundity (curve shifts to left) must be offset by a decrease in mortality (curve shifts to right). Inasmuch as whatever conditions adversely affect the birth rate are more likely to increase mortality than to decrease it, the population should decline to A. And inasmuch as whatever conditions enhance survivorship are more likely to enhance fecundity than to depress it, the population should increase to C. A steady-state population in other than space-limited populations (Fig. 3.1) seems improbable.

tion size. The first law of population dynamics shows us that steady-state populations have constant age-specific death rates and constant recruitment. Any increase (or decrease) in mortality must be balanced with a compensating increase (or decrease) in fecundity, if the population is to maintain a constant size. Food is unlikely to have this effect. As a population grows, per capita food consumption is certain to decline, sooner or later, if growth is not prevented by some other factor. Declining per capita food consumption will be accompanied by increasing mortality and decreasing fecundity (Fig. 3.12), with population decline rather than stability as a consequence.

The relationship between populations and their food supplies, as described above, cannot be fit to the requirements of the logistic equation, to the ideas of Lack (1954, 1966) that mortality rates balance birth rates, or to the ideas of Wynne-Edwards (1962) that birth rates balance mortality.

TIME AS A LIMITING FACTOR

Species with breeding and nonbreeding seasons have alternating periods of positive and negative growth rates, that is, alternating periods of increase and decline in numbers. Andrewartha and Birch (1954) suggested that the maximum population size is a function of (1) the initial size of the population, that is, the number of individuals that survive the preceding nonbreeding season, (2) the value of r, and (3) the length of the breeding season (Fig. 3.13). The first two may vary with density, but the length of the breeding season is almost certainly independent of density. If two populations have equal initial populations and values of r, they will differ in size if their breeding seasons differ in length.

The length of the nonbreeding season also can affect population size by determining the number of individuals that survive between breeding seasons. This number is a function of (1) the number of individuals (eggs,

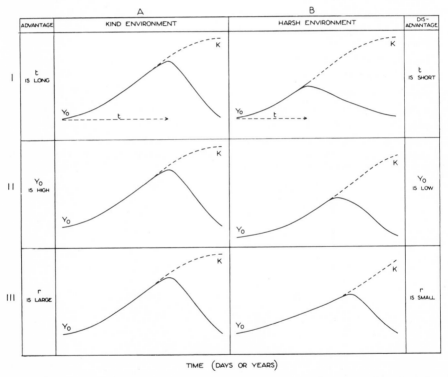

TIME (DAYS OR YEARS)

Figure 3.13. Three ways in which weather can affect a population's numbers. Periods favorable for growth alternate with unfavorable periods. (I) The favorable period is longer in the kind environment than in the harsh environment, and the population reaches a larger size in the kinder environment. (II) Survivorship during the unfavorable period is higher in the kind than in the harsh environment (see also Fig. 3.14), and thus the number (Y_0) entering the favorable period is greater in the kinder environment. (III) The population's growth rate (r) is greater in the kind than in the harsh environment, and thus in the same period of time the population in the kinder environment reaches a larger size (see text for fuller explanation). (From Andrewartha and Birch, 1954.)

larvae, young, or adults) that enter the nonbreeding season, (2) a survivorship schedule showing the probability of living through the nonbreeding season, and (3) the length of the nonbreeding season (Fig. 3.14). The first two may vary with population density, but the length of the nonbreeding season will be independent of density. Again, two populations of equal size, density, and survivorship at the beginning of the nonbreeding season will be different in size if their nonbreeding seasons are of unequal length.

Low-density populations should occur when breeding seasons are short and nonbreeding seasons are long, and high-density populations should occur when breeding seasons are long and nonbreeding seasons are short, other factors being equal. The fluctuations in the lengths of the breeding and nonbreeding seasons of a particular population will cause fluctuations in the population's size. Some of the fluctuations that occur in nature, then, may be the result of density-independent changes in the lengths of the breeding and nonbreeding seasons, perhaps caused by the weather.

Andrewartha and Birch (1954) considered weather a major influence on population numbers. Weather can affect numbers (1) by determining the length of the period favorable for breeding (I, Fig. 3.13), which we have seen determines the population's annual rate of increase, r'_a (Fig. 2.2 and Table 2.3), (2) by determining the size of the population at the beginning of the breeding season, a severe or prolonged nonbreeding season resulting in

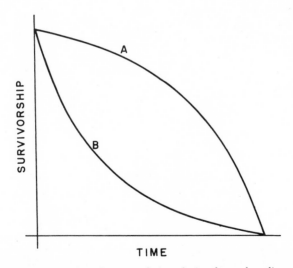

Figure 3.14. Survivorship of two populations during the nonbreeding season. The size of the population that survives the nonbreeding season depends on the survivorship (compare A and B), the initial size of the population (not shown), and the length of the nonbreeding season (compare either A or B with respect to time).

a smaller initial breeding population than a short or mild nonreproductive season (II, Fig. 3.13), and (3) by determining the probabilities of survivorship and reproduction and, therefore, the population's rate of increase r (III, Fig. 3.13).

Sorting out these multiple effects of weather on population numbers from each other and from other factors that influence numbers, such as resources, predators, and competitors, is no easy matter. It is necessary to have a thorough knowledge of the biology of the population, including measurements of population size, survivorship, and reproduction, and measurements of the relevant variables over a prolonged period of time.

Andrewartha and Birch (1954) reviewed the cases of four populations whose numbers they believed were largely determined by the weather. The most discussed case has been that of the apple blossom thrips (*Thrips imaginis*) of Australia (Davidson and Andrewartha, 1948a,b). Andrewartha and Birch (1954) summarize their thinking as follows:

> In our choice of independent variates for the regression we were guided by our knowledge of the biology of *Thrips imaginis* and the climate of the area where the population was living. In this climate, r for *T. imaginis* is likely to be negative at all seasons of the year except during spring and perhaps for a brief period during autumn in some years [Fig. 3.15]. The explanation for this may be given briefly as follows: During the summer, except for the early part, the places where pupae occur are likely to be so

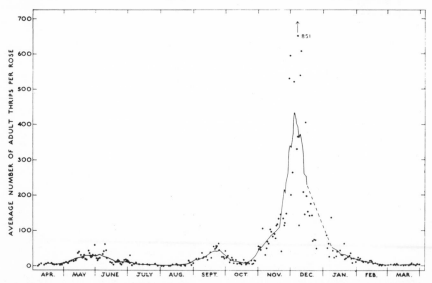

Figure 3.15. The numbers of *Thrips imaginis* per rose throughout 1 year, April 1932 to March 1933, in Australia. Each point represents the average number of *Thrips* per rose per day. The curve is a 15-day moving average of daily counts. (From Davidson and Andrewartha, 1948a.)

arid that few may be fortunate enough to survive; also, suitable flowers for breeding become scarce and sparsely scattered, and this adds to the hazards of life for adults, reducing their chances both of surviving and of leaving offspring behind them. During the winter, food continues to be sparsely distributed; also with the prevailing low temperature the thrips develop slowly, and many nymphs die simply because they fail to complete their development before the flowers in which they are living wither. With rising temperatures in the spring, the thrips develop more quickly; the soil, wet from winter and remoistened by spring rains, permits a high rate of survival among pupae. In some years flowers continue to be scarce at first, but it is a striking feature of this region that the blossoming of a high proportion of the natural vegetation is condensed into a brief period in the spring; the annuals, which preponderate, must set seeds before the summer drought sets in; and the perennials also tend to produce most of their flowers at this season. Consequently, the thrips flourish and multiply at this time until, as summer develops, the soil dries out and the flowers wither and disappear. The thrips which had been so abundant become few again.

Even when the thrips are most numerous, the flowers in which they are breeding do not appear to be overcrowded except perhaps locally or temporarily: while the thrips are multiplying, the flowers increase even more rapidly. Then, when the population begins to decline, the flowers become less crowded still. This may be partly because the survival-rate among pupae depends upon the moisture in the top half-inch of soil, whereas the plants draw their water from a greater depth. Also, as the flowers begin to thin out and the distances between suitable breeding places become greater, r may be reduced by this cause long before the flowers become absolutely scarce. . . .

Considerations of this sort led to the hypothesis that the numbers achieved by the thrips during each year were determined largely by the duration of the period that was favorable for their multiplication. When this period was prolonged, the thrips would ultimately reach higher numbers; when it was briefer, the decline would set in while the numbers were still relatively low.

The general argument, illustrated in Fig. 3.13, and the specific argument for *Thrips*, just quoted, seem reasonable. The mathematics of the dynamics of intermittently breeding populations (pp. 24–28) seems unchallengeable. Nevertheless, Smith (1961) reanalyzed the data of Davidson and Andrewartha, subjected the data to a different statistical treatment, challenged the conclusion of Andrewartha and Birch (1954) that "not only did we fail to find a 'density-dependent factor,' but we also showed there was no room for one," and argued that the numbers of *Thrips* were not determined by the weather but were regulated by density-dependent factors most of the time, although the density-dependent mechanism was not identified. Smith (1961) showed that *Thrips* population change was negatively correlated with its density and that the variance of the logarithm of average population size rapidly declined during the later portion of population increase, which Smith considered evidence in favor of density dependence.

The argument is not simple. Davidson and Andrewartha (1948b) originally argued that weather acted as a density-dependent factor, and Andrewartha and Birch (1954) themselves argued that "density-independent" factors did not exist. Because no component of the environment could be shown to act independently of density, there seemed no reason to single out particular factors and call them "density-dependent" (Andrewartha and Birch, 1954). When Andrewartha and Birch (1954) wrote that there was no room for a "density-dependent factor" in their analysis of the *Thrips* population, the "density-dependent factor" referred to was some form of "competition," which included intraspecific and interspecific competition for resources, predation, and parasites, as the term was defined and used by Nicholson (1933) and others. Thus, Andrewartha and Birch meant there was no room for "competition." Unfortunately, this semantic confusion was not cleared up in subsequent discussion (Andrewartha, 1963; Smith, 1963) and seems to have persisted to the present day.

Smith (1961) assumed that only density-dependent factors could account for the declining rate of population change with increasing density and for the decreasing variance with increasing density during the later stages of population growth. We have seen, however, that in space-limited populations (Fig. 3.1), the rate of population increase declines with increasing density even though no density-dependent factors are involved. Thus, a decreasing rate of growth with increasing density, a common observation (Tanner, 1966), does not necessarily imply that density-dependent factors are acting.

If the length of the reproductive season is limited by climatic conditions, what changes in growth rate should be expected? Conditions suitable for population growth do not turn on and off but change gradually. With the onset of suitable conditions at the beginning of the breeding season, we should expect r to be small. Then, r should rise to a maximum when conditions are most suitable for growth and decline through zero with the onset of unsuitable conditions at the end of the season (Fig. 3.16). Thus,

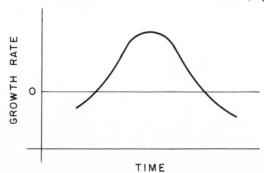

Figure 3.16. Change in the instantaneous rate of increase with time during the period favorable for growth. Breeding and nonbreeding seasons do not turn on and off. Instead, conditions gradually change from unfavorable to favorable to unfavorable again. Thus, r varies with favorability of breeding conditions.

the slower rate of growth need not reflect the effects of increasing density. It can result just as well from increasingly unsuitable climatic conditions.

Smith (1961) showed that the variance of the log of the average numbers of *Thrips* per rose per month increased strikingly in September and October and declined in November during the later stages of *Thrips* population growth (Fig. 3.17A). Smith (1961) contended that the low variance through most of the year showed that the population was usually regulated by density-dependent factors. Andrewartha (1963) suggested that changes in the variance in numbers reflected seasonal differences in the vagaries of the weather. The beginning of spring was less predictable than the beginning of summer. Smith (1963) responded that weather would affect the variance of r but not of numbers. In fact, the impressive increase in variance in September, October, and November reported by Smith (1961) when he analyzed the 81 consecutive months of data from April 1932 through December 1938 (Davidson and Andrewartha, 1948a) is not so impressive when only the first 72 months of data are used (Fig. 3.17B). Much of the large variances in September, October, and November can be attributed to the large numbers counted in those months in 1 year (1938) out of the 7.

I must support Davidson and Andrewartha (1948a,b) and Andrewartha and Birch (1954) in their interpretation of the factors affecting the numbers of *Thrips imaginis* in Australia. Clearly, weather can affect (1) the length of the breeding season, which in turn affects the annual rate of increase r'_a, (2) the initial population size, and (3) the instantaneous rate of increase r (Fig. 3.13). The effects of mild or severe weather on population size are well known to field ecologists, but the persisting belief in "density-dependent regulation" forces many to deny that weather can limit population size because weather is a "density-independent factor," in the sense that weather occurs independently of the population's density. Weather, however, is a "density-dependent factor," in the sense that the weather's action on population size is often density-dependent (Davidson and Andrewartha, 1948b; Andrewartha and Birch, 1954). To deny the limiting effects that weather has on the length of the breeding season and, therefore, on the population's numbers in favor of unidentified density-dependent regulating factors seems to place theory ahead of observation. The alternative theory of Andrewartha and Birch (1954) seems adequate.

The length of the breeding season, however, does not have the same limiting effect on population size as do space, crowding, and predation, discussed in previous sections. The length of the breeding season sets the size of a population in a particular season, depending on the given initial population and growth rate r. The length of the breeding season cannot limit the population over an extended number of seasons. Sooner or later a

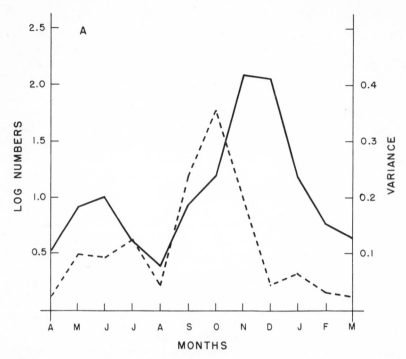

Figure 3.17. Monthly mean log numbers (solid line) and variance (dashed line) of *Thrips imaginis* in Australia. (A) Analysis of Smith (1961) based on data for 81 consecutive months, from April 1932 to December 1938, in Davidson and Andrewartha (1948a).

series of mild years can result in a large population, which then will be limited by space, crowding, or predation.

THE ROLE OF "DENSITY-DEPENDENT" FACTORS

In this chapter, we have considered the factors limiting the size of populations: space, food, predators, and time. In each of these, the limited population's rate of growth declines as its density increases, although in none of these was it necessary to resort to density-dependent factors to explain the decline in growth rate. The growth of such populations, then, is sigmoid. Thus, the fact that a particular population exhibits sigmoid growth does not constitute evidence that density-dependent factors are acting. Furthermore, the existence of density-dependent changes in mortality or fecundity, such as the decline in clutch size with increasing density in the Great Tit, already discussed (Fig. 1.8), does not indicate that such changes are "regulating" the population's size. An understanding of

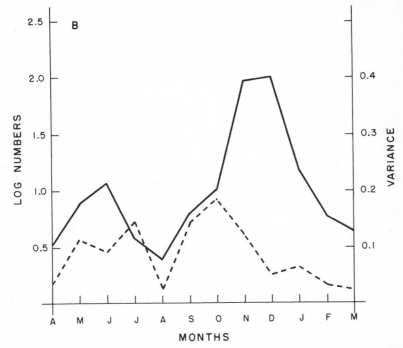

Figure 3.17. (B) The same analysis, but on the first 72 consecutive months, April 1932 to March 1938.

the dynamics of populations requires a close examination of the cause and effect relationships that are determining a population's birth and death rates.

In space-limited populations, the declining size of individual territories with increasing density may have no effect on either birth or death rates (Fig. 3.2). Yet as soon as the minimal territory size is reached, the population's growth rate declines with further population growth. This occurs because recruitment is limited and constant once the minimal territory size is reached. The critical factor, then, determining the maximum population size in space-limited populations is the minimum territory size, and this is a function of the behavioral and physiological attributes of the individuals of the population.

In time-limited populations, the weather affects the individuals' probabilities of surviving and reproducing and, therefore, the population's growth rate (r). The growth rate varies during the year (Fig. 3.16) such that during periods favorable for breeding the population's growth may be sigmoid. Density-dependent factors are not needed to explain the growth or cessation of growth of time-limited populations.

The declining birth rates and increasing death rates in food-limited and predator-limited populations also do not "regulate" population size. The factor that determines the size of food-limited populations is the amount of food available. The amount of food determines the point at which intraspecific competition results in reduced per capita food consumption and begins to reduce survivorship or fecundity or both (Fig. 3.5). The more food available in an area, the larger the size and the greater the density of the population when per capita food consumption declines (Fig. 3.5). Because every population must necessarily have a density and because decreasing fecundity and increasing mortality occur with increasing density, ecologists have concluded that the causal factor determining population size is "density-dependent." This is not a meaningful characterization for at least three reasons.

First, the term "density-dependent" is ambiguous because a density-dependent relationship can be either linear or curvilinear. The usual definitions of "density dependence" imply a linear relationship between birth and mortality rates and population density (see definitions in Chapter 1), although curvilinear relationships are often implied in the literature. The difference has important theoretical and practical consequences. The model proposed here for food-limited populations assumes that, below an upper critical density, per capita consumption of food is maximal and, therefore, there is no competition for food. Above the UCD, per capita consumption declines to the detriment of both survival and reproduction until birth and death rates are equal. In this model, per capita food consumption is a *consequence* of the total amount of food available rather than of the population's density because a population at the same density can have a different level of per capita food consumption if a different total amount of food were available. The independent variable is the food supply, which must be the *causal* factor in determining the size of food-limited populations. (It should be remembered that the total food supply does not refer to the total count of insects, for instance, actually occurring in a community. It refers to the number of insects that predators can capture and eat considering the morphological, physiological, and behavioral characteristics of both predators and prey. As stated earlier, I do not distinguish between absolute and relative abundance of food. What matters is the amount that consumers can obtain.)

In the logistic model of population growth there is no UCD. Per capita growth rate is a maximum at the lowest population density (Fig. 3.6) and is different at each density for populations exposed to different amounts of food. The cause and effect relationship between food supply and population density that can produce such a result has never been made clear.

Second, as can be seen in Fig. 3.5, the numerical value of the replacement and birth rates are the same regardless of the density at which per capita food consumption declines or at which each population stops growing. Low-density and high-density steady-state populations have the same replacement and birth rates. Therefore, replacement and birth rates of different populations are not correlated with density but with the degree of crowding as measured by per capita food consumption.

Furthermore, in the model presented here (Figs. 3.5–3.7) competition, mortality, and fecundity in a population vary only within the upper range of population densities (excepting the low birth rates at the lowest densities), there being a range of densities of variable extent through which birth, death, and therefore growth rates remain constant. Birth, death, and growth rates are not correlated with density, and therefore they should not be spoken of as density-dependent.

Third, three growing populations of different sizes were created by combining survivorship schedule A (Table 2.6), the replacement rate of which is 1.3, with a birth rate of 1.8, and by giving each population a different initial population. These populations grow at the same rate (Fig. 3.18), although different in size and density, and will continue to grow indefinitely unless some factor reduces the growth rate. The numerical value of the factor that is necessary to produce a steady-state condition is

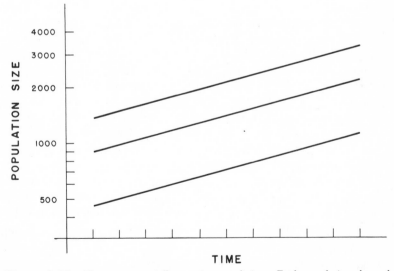

Figure 3.18. Three exponentially growing populations. Each population shown here has the same survivorship, age-specific birth rates, and stable age distribution. They grow at the same rate (r), Any factor or combination of factors that causes the population to reach a steady state ($r = 0$) must have the same value ($-r$) at whatever time or density it becomes effective.

precisely the same in each population at whatever time and density it acts and is, therefore, independent of the densities of the three populations.

The so-called density-dependent factors in population dynamics are really hypotheses that derive from a particular view of how populations function. In the alternative paradigm presented in this chapter, density-dependent factors play no role in the dynamics of populations.

According to the model a steady-state population cannot be established in food-limited or consumer-limited populations (Fig. 3.12). Consider a population that is at "steady state," that is, the birth rate balances the death rate. If the population is to remain in steady state, a decline in mortality must be accompanied by a decline in birth rate; if the birth rate increases, then mortality must increase. In nature, however, it is more likely that conditions leading to increased mortality should also depress fecundity, and those conditions leading to greater survivorship should also allow more successful reproduction. For example, by eating prey a predator reduces its food supply. A lower food supply is likely to reduce per capita food consumption, which should lower both survivorship and fecundity. Thus, a steady-state population is a virtual impossibility in food-limited and predator-limited populations. It is also impossible in time-limited populations. The space-limited population can maintain a steady state because territoriality can limit recruitment at some constant value. In the next chapters, "steady state" in quotation marks refers to the average population size about which numbers are fluctuating. When enclosed in quotation marks, "steady state" should not be taken to mean constant birth and death rates or stable age distributions.

DISTRIBUTION AND ABUNDANCE

Distribution and abundance are two sides of the same problem (Andrewartha and Birch, 1954), although they are usually treated separately. When they are considered together, we are immediately confronted with a paradox. It seems likely that survivorship and fecundity should vary with the quality of the environment, and intuitively we should expect that populations in suboptimal habitats would have higher mortality and lower fecundity than those living in optimal habitats. Yet the first law of population dynamics tells us that a high-density, steady-state population with high survivorship should have a *lower* birth rate than a low-density, low-survivorship, steady-state population in suboptimal habitat. This would be a peculiar state of affairs if it proved to be so. The conventional wisdom regarding population dynamics cannot help us. How is this riddle to be resolved?

For a population to exist at all, its birth rate must exceed its replacement rate, at least for some period of time. As we have already seen, birth and replacement rates vary with increasing intraspecific competition above the UCD in food-limited populations (Fig. 3.5), after all territories are occupied in space-limited populations (Fig. 3.2), and with the intensity of predation (Fig. 3.8; see also Fig. 5.3). Although each population's maximum birth rate and minimal replacement rate occur at low densities, a population living under optimal conditions will have a greater maximum birth rate and a smaller minimal replacement rate than a population living under suboptimal conditions (Fig. 3.19A). As each population grows in size, it eventually reaches a point at which per capita food consumption declines, territorial space becomes unavailable, or predation intensity increases, bringing about a reduction in birth rates and an increase in mortality. When the birth rate falls below the replacement rate, the population stops growing. Each population will stop growing at and fluctuate about a different mean density, the greatest density occurring in optimal habitat (Fig. 3.19B). Even though each population has different birth and replacement rates (and therefore growth rates) at the lower population densities when all populations are growing, intraspecific competition at the higher densities decrease birth rates and increase replacement rates until they are about the same in all steady-state populations. High-density, steady-state populations in optimal habitat have the same birth and replacement rates as do low-density, steady-state populations in suboptimal habitat. Populations living in optimal habitat reach the greatest density because they have the greatest growth rate and because they grow for the longest period of time.

Recall, however, that survivorship and reproductive schedules for populations, rather than for cohorts, include emigration as deaths and immigration as recruitment. Therefore, an increasing replacement rate reflects increasing mortality, increasing emigration, or both, and, conversely, a decrease in the replacement rate reflects decreasing mortality, increasing immigration, or both. Thus, as per capita food supplies decline or available territorial spaces decline, the increasing replacement rate may result from increasing dispersal rather than from increasing mortality, although probably both are occurring. Low replacement rates of populations in suboptimal habitats may mean that immigration is high rather than mortality low. Thus, although the population *statistics* (birth and replacement rates, age distributions, etc.) may be the same for "steady-state" populations in optimal and suboptimal habitats, the *dynamics* of the populations is different. The greater survivorship of individuals born into or living in optimal habitat is offset by emigration, which is counted as deaths, and the poorer survivorship of individuals born into and living in suboptimal habitats is offset by immigration from optimal habitats, the immigrants replacing the

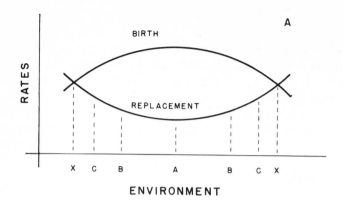

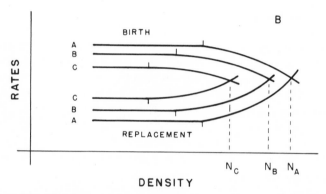

Figure 3.19. Distribution and abundance of species. (A) Local populations of the same species will be characterized by different maximum birth rates and minimum replacement rates because they occupy areas of different quality with respect to food supply, presence of predators, and physical factors affecting birth and death rates. By definition, the optimal environment (A) allows the greatest birth rate and lowest death rate compared with suboptimal habitats (B and C). Environment X marks the edge of the species' range. (B) The growth of local populations in different environments (A, B, C). The population in the optimal environment has the greatest growth rate and longest period of growth, and, therefore, it reaches the greatest density.

deaths and reducing the apparent age-specific death rates. Identity of survivorship and reproductive schedules or populations living in optimal and suboptimal habitats can hardly be expected, but they will be more similar than might at first be expected.

In order to study distribution and abundance of animals, Andrewartha and Birch (1954) divided the environmental factors affecting survivorship and reproduction into four components: (1) weather, (2) food, (3) other animals, and (4) a place in which to live. Other animals include those of the same kind (intraspecific competition) as well as interspecific competitors, predators, and disease organisms. Each of these components varies geographically, and thus population densities will vary geographically. Specifically, with respect to the models presented in this chapter, all four components do affect survivorship and reproduction, but each varies in effect in time and space.

Weather affects survivorship and reproduction at all times. Its effects on numbers are determining the length of the season favorable for reproduction (I, Fig. 3.13) and the population's maximal potential growth rate (III, Fig. 3.13 and 3.19), both especially important in determining the coefficient of annual increase r'_a (Fig. 2.1; Table 2.3). The amount of available food has no effect on either population numbers or growth rate until the UCD is reached, after which increasing intraspecific competition for food increases mortality and reduces fecundity (Fig. 3.5). Other animals of the same species have no effect on either population numbers or growth rate until the UCD is reached in food-limited populations (Fig. 3.5) or all available territory spaces are occupied in space-limited populations (Figs. 3.1 and 3.2). Any predation or disease reduces a population's growth rate and "steady-state" size (Fig. 3.8, but see more detailed treatment in Chapter 5). The effects of interspecific competition require separate treatment in Chapter 6. A place in which to live affects survivorship and reproduction in its capacity to provide protection from weather or predators.

Research on populations has been piecemeal. The complexity of the interrelations of individuals with each other and with their environments almost precludes a holistic approach. Nevertheless, the collection of counts, either in space or in time, and the fitting of mathematical equations to the counts do not seem suitable substitutes for discovering how the four components of the environment affect the probabilities of survival and reproduction.

EVALUATION OF OLDER THEORIES

As I reread the older literature on population dynamics, I become more convinced that earlier theorists had in their minds models similar to those presented in this chapter, but words failed them. Each author selected terms that meant different things to different readers, and this ambiguity coupled with each author's emphasis on different aspects of the population problem led to the impasse of the 1950s.

Perhaps the most explicit authors were Andrewartha and Birch (1954), who summed up their theory as follows (pp. 660–661): "The number of animals in a natural population may be limited in three ways: (a) by shortage of material resources, such as food, places in which to make nests, etc.; (b) by inaccessibility of these material resources relative to the animals' capacities for dispersal and searching; and (c) by shortage of time when the rate of increase r is positive. Of these ways, the first is probably the least, and the last is probably the most, important in nature."

I completely agree, except this is not enough. There is no mechanism given to explain how limited resources can bring about a zero growth rate without intraspecific competition, a phenomenon that Andrewartha and Birch strongly rejected as having any importance in the determination of a population's numbers. The authors' stand on competition, their insistence that density-dependent factors were unnecessary for limiting the size of populations, and their diagrammatic, nonmathematical explanations of population limitation (Fig. 3.13) were not sufficiently convincing to other ecologists, who found greater security with mathematical equations, which seemed to provide a sounder basis for prediction and experimentation. Andrewartha and Birch were correct in their evaluation of the role of competition in determining the numbers of a population but erred in concluding that competition, induced by the limiting factor, was not the mechanism causing decreased survivorship and birth rate. With this exception, I believe the views of Andrewartha and Birch on how the numbers of populations are determined are compatible with my own.

Curiously, stripped of its unfortunate terminology, the theory of Nicholson, the chief antagonist of Andrewartha and Birch, is not unpalatable to me. Nicholson (1958b) briefly outlined his theory of "self-regulation": "If a population is reduced to a very low level, due to some adverse factor (which in temperate regions may be no more than a change in seasons), the population must be expected to multiply without check for some time after favorable conditions return; opposition to population growth only begins when the population becomes sufficiently high to cause significant depletion [of one or more requisites], or some other density-governing reaction [e.g., increasing predation]." By deleting the term "density-governing" we have a statement that is virtually identical with one that I could have made.

Furthermore, a supporter of Nicholson's ideas, Solomon (1957), wrote, "We should not assume that a population is necessarily in the process of being regulated all the time. Some populations may be so, others seem not to be. For the looser sort of control, it is necessary only that one density-dependent process or another should come into effective operation whenever the density becomes high, the critical density value tending to vary continually with environmental conditions. The density may remain below this level for considerable periods because of unfavorable weather etc., and during these periods there may be no factor operating with a density-dependence of much significance." Again, if we delete the term "density-dependent," the general idea bears a resemblance to my own (and even to that of Andrewartha and Birch).

What about these "density-dependent" factors? According to Solomon (1964), density-dependent factors provide the "finishing touches" to population regulation, acting perhaps on only 2% of the original population, 98% of which was eliminated by density-dependent mortality.

It seems clear to me that if "opposition to population growth only begins when the population becomes sufficiently high," and if "it is necessary only that one density-dependent process or another should come into effective operation whenever the density becomes high," and if density-dependent factors act on as little as 2% of the population, then the so-called density-dependent factors are not correlated with a population's density. They are correlated with density only above some threshold level, there being a more or less wide range of densities of a population when the so-called density-dependent factors are not affecting either mortality or fecundity but the populations are fluctuating in size. Such a conception of density-dependence seems to be inconsistent with Solomon's (1958a) definition for "directly density-related factor," a virtual synonym for what many refer to as a "density-dependent factor": a factor "showing a positive correlation between adverse action on population growth and density."

I find it difficult to distinguish the conceptions that gave rise to the Nicholson and Solomon quotations from the conception that is the source of the following quotation from Ehrlich and Birch (1967): "We are not denying that numbers of some populations may be influenced by so-called 'regulatory factors,' i.e., whose depressive effect on rate of increase is positively correlated with density (Solomon, 1964). We would deny that there is any convincing evidence that the numbers of all populations are primarily determined by density regulating factors." As I stated earlier, my intention in this book is not to sort out the ideas of others, if only because those ideas were put so ambiguously, in my opinion, that they have been and can be interpreted, or misinterpreted, in so many ways.

Outside the semantic difficulties remains the problem of how often a population's numbers are limited by resources (food and space), how often by predators or disease, and how often by the length of time the population's growth rate is positive. This was the fundamental difference between the ideas of Nicholson and the ideas of Andrewartha and Birch. This is not a problem that can be resolved by appealing to theory. It requires a great deal of further research in the field.

4

POPULATION DYNAMICS AND NATURAL SELECTION

The evolution and dynamics of populations are inextricably intertwined. Shifting gene frequencies can change the l_x and m_x schedules, and changes in population size can act as a selective agent because the advantage of a genotype may vary with population density. This, of course, was the basis for the concepts of r-selection and K-selection, which seemed to bridge the gap between population dynamics and evolution. Hairston *et al.* (1970) believed that r-selection and K-selection confused natural selection and population dynamics, but there are other sources of confusion in this area of population biology.

One cause of confusion is the use of symbols that have several meanings. For example, l_x has been used to mean either the proportion of a cohort that reaches age x or the probability of an individual reaching age x. The difference seems subtle because an individual's probability of reaching age x can be determined only by observing a cohort of individuals throughout the course of its existence. Also, m_x can mean either the birth rate of age class x or the probable number of offspring born to an individual of age x. Again, the difference seems subtle. But l_x and m_x are usually determined from, and applied to, genetically diverse populations. In such populations, genetically different individuals are likely to have different probabilities of surviving and reproducing. If they do not, natural selection cannot occur.

Theoretically, each individual has probability schedules of survivorship and fecundity, if they could be known, depending on its genetic inheritance. Assuming a particular set of values for a given genotype, we can use Eq. (2.4) to calculate the genotype's r, the equivalent of Fisher's (1930) Malthusean parameter, m. In this special case, r is a measure of a genotype's potential success relative to other genotypes with different survivorship and fecundity schedules. However, individuals with particular genotypes do not accumulate, that is, change in number or frequency, except by asexual reproduction and parthenogenesis. What does change over the course of time is the frequency of a particular trait, and this usually is what we are interested in.

I propose to retain l_x and m_x to represent the average values of survivorship and reproduction of individuals in genetically diverse populations for purposes of population dynamics and use λ_x to represent the probability of an individual with a particular trait to reach age x and μ_x to represent the probable number of offspring that an individual with the same trait will either sire or bear at age x. We can also replace r with ρ, the rate of increase of a population of individuals with the trait of interest. Thus, for calculating the rates of increase of particular traits, Lotka's equation [Eq. (2.4)] becomes

$$1 = \Sigma \lambda_x \mu_x e^{-\rho x}. \tag{4.1}$$

Distinguishing l_x from λ_x, m_x from μ_x, and r from ρ is useful. We can now think of a genetically diverse population with characteristic l_x, m_x, and r values as consisting of many subpopulations, each with characteristic λ_x, μ_x, and ρ values. If differences in λ_x and μ_x values have a genetic basis, the traits of the individuals with the greatest ρ will increase in frequency compared with traits of individuals with lower ρ. They will increase in frequency whether the population is increasing, decreasing, or remaining constant in size.

Another parameter of interest is Fisher's (1930) reproductive value v_x, which is a measure of the expected number of future offspring for an individual of age x compared with its expected number of offspring at birth (v_0). Setting v_0 to 1 and using the symbols introduced above, the reproductive value at age x of individuals with a given trait is given by (Fisher, 1930):

$$v_x = \frac{e^{\rho x}}{\lambda_x} \sum_{t=x}^{\infty} \lambda_t \mu_t e^{-\rho t} \tag{4.2}$$

The reproductive value increases from birth, reaches a maximum, and declines to zero at the end of the reproductive period (Fig. 4.1). The value of $\Sigma \lambda_t \mu_t e^{-\rho t}$ is a maximum at the age of first reproduction and decreases with increasing age. Prior to the age of first reproduction, $\Sigma \lambda_t \mu_t e^{-\rho t}$ is a constant (specifically, 1.000). The value of $e^{\rho x}/\lambda_x$ increases with increasing

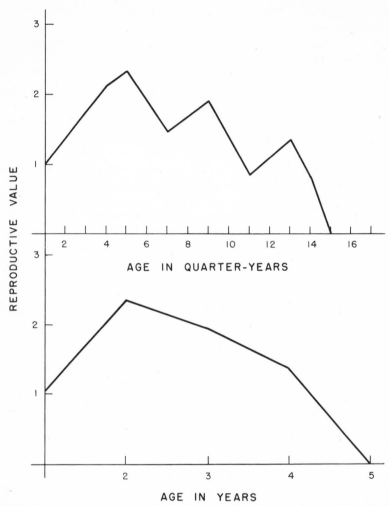

Figure 4.1. Comparison of the reproductive values for age classes of the identical population when data are collected each quarter-year and annually.

age. Whether the age function of reproductive value following the age of first reproduction is smooth or saw-toothed (see Fig. 4.1) depends on the time interval chosen for the age classes (Pianka and Parker, 1975). Fisher's concept of reproductive value with age is consistent with intuitive notions. If we were to buy an animal for breeding purposes, we would expect to pay more for a recently matured one because younger animals would still have a chance of dying before reaching maturity and older animals have less time in which to reproduce.

The concept of reproductive value has practical application. Colonies established by newly matured individuals are likely to be more successful than those established by an equal number of immature or an equal number of older animals (MacArthur and Wilson, 1967; Mertz, 1970). Also, a predator hunting younger prey will lower the sustainable yield compared with predators hunting older prey (see Chapter 5).

It is worth noting that ρ and v_x are derived values. They are consequences of λ_x and μ_x schedules. Natural selection acts on traits that affect survivorship and reproduction, and those that maximize ρ will be favored in genetically diverse populations. Selection will also maximize v_x at every age (Taylor *et al.*, 1974), but models or discussions concerned with maximizing ρ, v_x, or some other measure of fitness will always be incomplete. We must next ask whether maximum ρ and v_x have come about because of a change in survivorship or reproduction or both. We must return to the original problems; e.g., why is a given clutch size rather than another favored under these conditions, why do individuals begin breeding at one age instead of another, and why are some populations characterized by semelparity and others by iteroparity?

Fisher (1930) also suggested that "it would be instructive to know not only by what physiological mechanism a just apportionment is made between the nutriment devoted to the gonads and that devoted to the rest of the parental organism, but also what circumstances in the life-history and environment would render profitable the diversion of a greater or lesser share of the available resources towards reproduction" (p. 47). This has led to the concept of "reproductive effort" (Williams, 1966a,b; Tinkle, 1969). Although all resources are devoted to reproduction (present reproduction plus future reproduction as measured by survivorship), the problem posed by Fisher is an important one. What are the physiological processes by which resources are allocated to present and future reproduction? The danger is in making *a priori* evaluations of what constitutes a greater or lesser amount of reproductive effort. For example, Williams (1966a) and Tinkle (1969) considered the laying of several clutches per season a greater reproductive effort than laying a single clutch, whereas Spencer and Steinhoff (1968) assumed the reverse. It seems more probable that whether females lay a single clutch or several clutches per season will depend on the advantage each alternative has in particular environmental conditions. Under some conditions, a single clutch could demand greater effort, whereas under other conditions multiple clutches could demand greater effort. Again, we are likely to be more successful in modeling the evolution of reproductive patterns by asking how natural selection has shaped the probability schedules of survival (λ_x) and reproduction (μ_x).

A major source of confusion seems to be how to apply the scientific method to ecological problems. Stearns (1976) reviewed the literature pertaining to the evolution of life history patterns, and he was disturbed by the preponderance of papers that made plausible but "absolutely unfalsifiable" predictions, papers that "disregard a consideration of hard evidence in preference for a discussion of ideas for their own sake," and "a confusion of untested ideas which are judged, not on their ability to withstand empirical tests, but on the difficulty of the mathematics used or the obscurity of the theoretical development." This trend does not constitute science as Stearns understands it, and I agree completely with his remarks.

The evolutionary biologist must keep distinct not only the difference between population parameters and individual attributes but (1) the purely descriptive relationships between the various parameters, (2) the evolution of individual characteristics relating to survival and reproduction, and (3) the consequences for population growth of the evolution of increased (or decreased) survival or reproduction. Let us take up these points one at a time.

Description

The mathematical equations describing the relationships between population parameters (l_x, m_x, r, b, d, R, T, and c_x) were developed by Lotka and were presented in Chapter 2. These equations have been and can be used to discuss the consequences of changing the probabilities of survival and reproduction of individuals, but as I indicated above I believe communication is clarified by substituting λ_x for l_x, μ_x for m_x, and ρ for r. The parameter ρ is the rate of increase of a subpopulation of individuals whose probabilities of survival and reproduction are λ_x and μ_x, respectively. This same subpopulation will have characteristic values for b, d, R_0, T, and c_x, if the selective environment has been unchanged for sufficient time, and if one were interested in calculating them.

In his important and stimulating paper, Cole (1954) examined how variations in one or more parameters affected a population's rate of growth (r). Most of Cole's results are straightforward. The population's rate of growth can be increased by increasing individual life expectancy, increasing clutch or litter size, and increasing the individual's total reproduction by decreasing the age of first reproduction, that is, shortening the generation time. Less straightforward is the question of the evolution of iteroparity (multiple reproduction) (Cole, 1954) because "for an annual species, the absolute gain in intrinsic population growth which could be achieved by changing to the perennial reproductive habit would be exactly equivalent to adding one individual to the average litter size" (p. 118).

How natural selection leads to the evolution of iteroparity and other life history attributes cannot be discovered by manipulating mathematical equations. One might guess that natural selection tends to increase life expectancy and clutch or litter size and decrease the age of first reproduction because each of these changes results in a greater ρ, if the values of the other parameters are unchanged. However, species exist whose members have short life expectancies, or small clutch or litter sizes, or long periods of deferred maturity. Evidently, natural selection does not always result in long life expectancies, large clutch or litter sizes, or early maturity. In the final paragraph of his famous paper, Cole (1954) wrote, "The number of conceivable life-history patterns is essentially infinite, if we judge by the possible combinations of the individual features that have been observed. Every existing pattern may be presumed to have survival value under certain environmental conditions" (p. 135). The problem is to determine which environmental conditions lead to particular life history patterns.

Evolution of Life History Patterns

In this section, I will first propose a biological, cause and effect sequence of evolutionary events leading to observed life history patterns, keeping the discussion as free as possible from mention of already proposed theories in order to present a coherent thesis. The method is to establish a set of assumptions stated in such a way that it provides a prediction of the life history pattern to be expected under particular environmental conditions. Predictions are then compared with available empirical evidence.

Stearns (1977) discussed the theoretical and empirical sources of ambiguity in the various models that have been proposed to account for the evolution of life history parameters. Stearns (1977) believed, "These ambiguities all share a general form. Each represents an unanalyzed complexity or subtlety, and for each we do not know whether explicit consideration of the problem would make any difference to our predictions" (p. 146). Some of these ambiguities cannot be directly tested. We are never likely to know the probabilities of survival and reproduction of an individual with a particular genotype. Indeed (Fisher, 1930), "the actuarial information necessary for the calculation of the genetic changes actually in progress in a population of organisms, will always be lacking: if only because the number of different genotypes for each of which the Malthusian parameter is required will often, perhaps always, exceed the number of organisms in the population" (p. 47). Such deficiencies in knowledge should not deter the theorist any more than ignorance of Mendelian genetics prevented Darwin from developing his theory of natural selection. The theorist must content himself with creating a theory that makes few assumptions but many predictions that are consistent with the present state of empirical knowledge.

Following the presentation of a new theory that predicts patterns of life history parameters, I discuss earlier theories and, finally, the concepts of r-selection and K-selection.

A THEORY OF EVOLUTION OF LIFE HISTORY PATTERNS

Assumption 1. In a genetically diverse population, natural selection increases the frequencies of the traits of those individuals whose probabilities of survival and reproduction result in the greatest rate of increase, ρ, under prevailing conditions. In almost all cases the individuals with the maximal ρ will also have the maximal $\Sigma \lambda_x \mu_x$, and, because the exceptional cases are unlikely alternative life histories within a population, the following discussion of the evolution of life history characteristics is put in terms of selecting for the greatest $\Sigma \lambda_x \mu_x$.

We can examine the relationship between $\Sigma \lambda_x \mu_x$ and ρ by rearranging Eq. (2.8) and substituting ρ for r,

$$\ln R_0 = \rho T,$$

and substituting $\Sigma \lambda_x \mu_x$ for R_0,

$$\ln \Sigma \lambda_x \mu_x = \rho T.$$

An increase in λ_x or μ_x at any age will increase $\Sigma \lambda_x \mu_x$ and will increase or decrease T, the generation time, depending on whether the change occurs early or late in the life cycle, but the change in T (i.e., $\Sigma x \lambda_x \mu_x / \Sigma \lambda_x \mu_x$) will be small relative to the change in $\ln \Sigma \lambda_x \mu_x$. Thus, ρ should increase when λ_x or μ_x increases. Increasing ρ without also increasing $\Sigma \lambda_x \mu_x$ requires decreasing T without increasing $\Sigma \lambda_x \mu_x$. The alternative life history must not only increase fecundity of the younger age classes but reduce fecundity or life expectancy of older age classes. The other exceptional case, increasing $\Sigma \lambda_x \mu_x$ without increasing ρ, requires increasing T by postponing reproduction to a late age and considerably increasing fecundity of the reproductive ages. The shift from a particular life history to one in which an increase in $\Sigma \lambda_x \mu_x$ does not coincide with an increase in ρ, or vice versa, requires larger multiple changes in survivorship and reproductive schedules than I consider plausible.

Assumption 2. Natural selection always maximizes life expectancy, that is, increases the probabilities of surviving to each age x. Therefore, the probability of surviving to age x (λ_x) is the best that can be expected under prevailing conditions. There is no selection for shortening life expectancy, but no individual can be immortal. There is always some probability of being killed by a predator, by disease, by accident, or by starvation, and there is always a cost in survivorship associated with reproduction.

This assumption does not preclude the evolution of senescence (Williams, 1957; Hamilton, 1966). Senescence is a consequence of the eventual inevitable decline in the probability of breeding with age (Williams, 1957). A gene that leads to a decline in life expectancy cannot be selected because it would reduce $\Sigma \lambda_x \mu_x$ relative to that of other members of the population without that gene. Physiological death will always follow the age at which further reproduction is improbable.

Assumption 3. Each additional egg in a clutch or young in a litter results in a reduction of life expectancy of females (Fig. 4.2) and, perhaps,

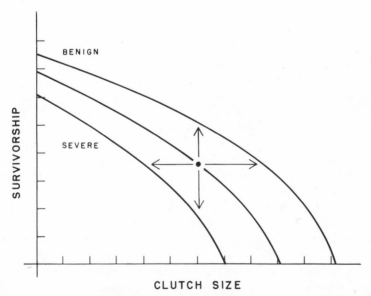

Figure 4.2. Relationship between clutch size and annual survivorship. As clutch size increases, the probability of surviving to the next breeding season declines, regardless of environmental conditions. A particular clutch size, say six, becomes more costly as conditions worsen. In a fluctuating environment, a female can either lay the same number of eggs with varying cost in survivorship (vertical arrows) or lay a different number of eggs with no change in cost in survivorship (horizontal arrows).

TABLE 4.1

Studies Showing Costs of Reproduction

Species	Reference
I. Virgin females live longer than reproducing females	
Cockroach (*Periplaneta americana*)	Griffiths and Tauber (1942)
Milkweed bug (*Oncopeltus fasciatus*)	Dingle (1968)
Carabid beetles (*Agonum* spp.)	Murdoch (1966)
Sheep blowfly (*Lucilia cuprina*)	Nicholson (1954a,b)
Drosophila subobscura	Maynard Smith (1959)
Daphnia pulex	Frank *et al.* (1957)
Daphnia obtusa	Slobodkin (1959)
Minnow (*Pimephales promelus*)	Markus (1934)
II. Mortality increases during reproductive season	
Gastropods (*Shaskyus* sp. and *Ocenebra* sp.)	Fotheringham (1971)
House Sparrow (*Passer domesticus*)	Summers-Smith (1956)
European Blackbird (*Turdus merula*)	Snow (1958)
Olive baboon (*Papio anubis*)	Berger (1972)

of males, especially in those species in which males establish territories and care for the young. A greater mortality of breeders compared with nonbreeders has been shown in a variety of invertebrates and vertebrates (Table 4.1). The data presented in the papers cited in Table 4.1 show only that reproduction has a cost in further life expectancy. The costs of producing and caring for each egg or young have not been measured.

Assumption 4. The clutch or litter represents the *fewest* eggs or young that a female with given probabilities of survival (λ_x) can produce during each reproductive effort and still be able to replace herself and her mate in the next generation. The benefit of an additional egg is never worth the cost in decreasing survivorship.

Some readers may believe that additional eggs can more than offset the cost of lowering survivorship, contrary to this assumption. Some may believe, furthermore, that the gain from greater fecundity may even offset the cost of dying, leading to an early death, contrary to assumptions 1 and 2. These notions, however, are *ad hoc* hypotheses that constitute assumptions of alternative models, which themselves must be evaluated by comparing the predictions that follow from their assumptions with available observations. In the discussion that follows, I will evaluate a model that includes the assumption that an additional egg or young is not worth the cost in additional mortality to either the young or the parents.

In order to illustrate the consequences of this set of assumptions, consider a hypothetical, numerical example of a species with determinate growth and constant fecundity with age. The annual replacement egg or young production μ is given by

$$\mu = 2/ \sum_{\alpha}^{\omega} \lambda_x \qquad (4.3)$$

when α is the age of first reproduction and ω is the age of last reproduction. Given the probabilities of survival (λ_x) in Table 4.2, the replacement rate is easily calculated, from Eq. (2.13), as 0.727. This means that the female must produce 1.455 young each year of her life, including her first, in order to replace both herself and her mate. If breeding is postponed until the second year, the female must produce 5.333 young during each year of her reproductive life. If breeding is postponed further, annual reproduction must be even greater (Table 4.2). Because eggs and young occur in increments of 1, the annual production must be 2, 6, . . ., 80, respectively, per year of reproductive life. Which life history will predominate depends on physiological and environmental constraints on reproduction. There is, no doubt, a minimal age at which reproduction can occur with any probability of being successful, a maximal number of eggs a female can lay during a single reproductive effort, and a maximal number of young that can be provided for in species with parental behavior. If breeding is certainly unsuccessful in the first year and if the maximal number of eggs that can be laid and young provided for is 10 in the present example (Table 4.2), then breeding had better begin in the second year, in which case the clutch size would be six. A smaller clutch size results in $\Sigma \lambda_x \mu_x < 2$ and a failure of those females with smaller clutches to replace themselves and their mates. A clutch size larger than six does not result in greater lifetime productivity ($\Sigma \lambda_x \mu_x$) because survivorship is reduced by the increased reproductive effort (assumption 3).

TABLE 4.2

Effect of Age of First Breeding on Clutch Size

Age x	Probability of surviving to age x	Age at first reproduction	Average m_x	Minimum number of eggs at each age
1	1.000	1	1.455	2
2	0.200	2	5.333	6
3	0.100	3	11.429	12
4	0.050	4	26.667	27
5	0.025	5	80.000	80
$b_r = 0.727$				

Now, suppose that a mutation occurs that leads to greater crypticity. It should increase in frequency if the mutants have a lower probability of being killed by a predator and, therefore, a higher probability of surviving than the individuals who do not carry this gene. The mutants' replacement clutch size is now less than 5.3. If the gain in longevity from greater crypticity reduces the replacement clutch to, say, 5.1, the females will still lay a six-egg clutch, but, if the gain reduces the replacement clutch to 5.0 or less, the females should lay a five-egg clutch.

The problem of determining the replacement clutch for species with indeterminate growth is more complex because annual fecundity increases with age as individuals grow larger. Equation (4.3) is, therefore, inapplicable. However, for a given survivorship schedule and age of first reproduction, the initial annual egg or young production will be less than expected from Eq. (4.3), and the annual egg or young production at the final age of reproduction will be larger than expected from Eq. (4.3). No specific numbers can be predicted because the number of permutations is large.

This is the basic model for the prediction of life history patterns. We must now apply this model to specific situations, predict how life history patterns should vary with environmental conditions, and compare the predictions with empirical evidence.

Clutch Size

The clutch or litter size of an animal is the number of eggs laid or young born to a female during a single reproductive effort. It is not equivalent to m_x, which is the age-specific birth rate and represents the average number of eggs laid or young born to females of age x (or, as in this book, to all individuals including males of age x), including those that do not breed (clutch size 0). The parameter μ_x is the expected number of eggs or young sired by or born to an individual of age x. These eggs or young may be distributed in a single clutch or litter or among several during the time interval x, $x + 1$.

Relationship between Survivorship and Fecundity

We should expect an inverse relationship between survivorship and fecundity. A long-lived individual will usually have a greater $\Sigma_\alpha^\omega \lambda_x$ than a short-lived one. Thus, from Eq. (4.3) the replacement clutch of a long-lived organism will be smaller than that of a short-lived one. The only exceptions are those species whose members are long-lived but defer breeding to a later age than other species, reducing $\Sigma_\alpha^\omega \lambda_x$.

An inverse relationship between survivorship and annual fecundity has been shown in lizards (Fig. 4.3), is well known in birds (Lack, 1954, 1966,

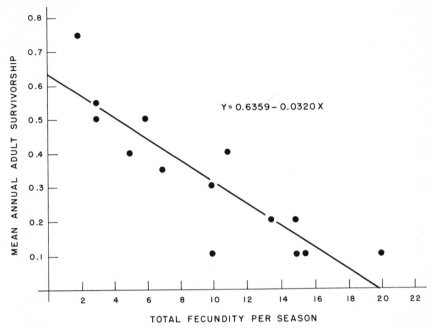

Figure 4.3. Relationship between total fecundity per reproductive season and probability of surviving to a later breeding season in lizards. (From Tinkle, 1969.)

1968), and is one of the correlates in the comparison of so-called r-selected and K-selected species (Table 4.3).

Field data from "steady-state" populations will always show an inverse relationship between survivorship (l_x) and fecundity (m_x). This may come about, as suggested by Lack (1954, 1966), by parents raising as many young as they can nourish, with density-dependent mortality eliminating the excess. Thus, these data do not constitute a critical set that distinguishes one theory as better than another. That evaluation must await the full analysis.

Geographic Variation

Two constraints on reproduction are the length of the breeding season and the length of time the young require to develop to a stage that either can survive the nonbreeding season or is independent of the parents. If development time is long relative to the length of the breeding season, individuals should have a single, large clutch. If development time is short relative to the length of the breeding season, the possibility of producing several clutches per season arises (Fig. 4.4). If two populations comprise individuals with the same probability schedule of survival, we should

TABLE 4.3

Correlates of *r*- and *K*-Selection[a]

Condition	*r*-Selection	*K*-Selection
Climate	Variable and/or unpredictable; uncertain	Fairly constant and/or predictable; more certain
Mortality	Often catastrophic, non-directed, density independent	More directed, density dependent
Survivorship	Often type III (Deevey, 1947)	Usually types I and II (Deevey, 1947)
Population size	Variable in time, nonequilibrium; usually well below carrying capacity of environment; unsaturated communities or portions thereof; ecological vacuums; recolonization each year	Fairly constant in time, equilibrium; at or near carrying capacity of the environment; saturated communities; no recolonization necessary
Intra- and interspecific competition	Variable, often lax	Usually keen
Relative abundance	Often does not fit MacArthur's broken-stick model (King, 1964)	Frequently fits MacArthur's broken-stick model (King, 1964)
Selection favors	1. Rapid development 2. High r_{max} 3. Early reproduction 4. Small body size 5. Semelparity; single reproduction	1. Slower development; greater competitive ability 2. Lower resource thresholds 3. Delayed reproduction 4. Larger body size 5. Iteroparity; repeated reproduction
Length of life	Short, usually less than 1 year	Longer, usually more than 1 year
Emphasis in energy utilization	Productivity	Efficiency
Colonizing ability	Large	Small
Social behavior	Weak; mostly schools, herds, aggregations	Frequently well developed

[a] From Pianka (1970) and Wilson (1975).

expect the one living in a region with long breeding seasons with respect to development time to have several small clutches and the one living in a region with a short breeding season with respect to development time to have a single, large clutch (Spencer and Steinhoff, 1968). In the hypothetical, numerical example (Table 4.2), with reproduction beginning in the second year, a female can replace herself and her mate if she lays one

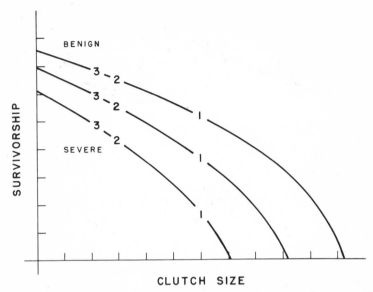

Figure 4.4. Relationship between clutch size and annual survivorship, as modified by the length of the breeding season. If the total annual replacement number of eggs is six, a female can lay a single clutch of six eggs, two clutches of three eggs, or three clutches of two eggs. How many clutches are laid depends on the relative lengths of the development time of the young to independence and the time available for successful reproduction. A short development time and long breeding season should result in several small clutches. A long development time and short breeding season should result in a single large clutch.

clutch of six eggs, or two clutches of three eggs, or three clutches of two eggs (Fig. 4.4). How many clutches of how many eggs a female lays depends on the length of the breeding season. If a female lays as few eggs as possible at each reproductive effort, a long breeding season allows production of several small clutches.

The most obvious determinant of the length of the breeding season is the prevailing weather conditions, which vary geographically. In general, periods with conditions suitable for breeding of many plants and animals become shorter from tropical latitudes toward the poles, from maritime to continental climates, and from low to high elevations.

An increasing clutch size with increasing latitude is well known and well documented in birds (Lack, 1947a, 1948a, 1954, 1966, 1968; Cody, 1966, 1971; Klomp, 1970; von Haartman, 1971). A few examples are shown in Figs. 4.5 and 4.6. Not all species or groups, however, show this trend. For example, all species of Procellariiformes have one-egg clutches, and all hummingbirds and most swifts and pigeons have two-egg clutches. Two European passerines have a reverse trend, larger clutches at more south-

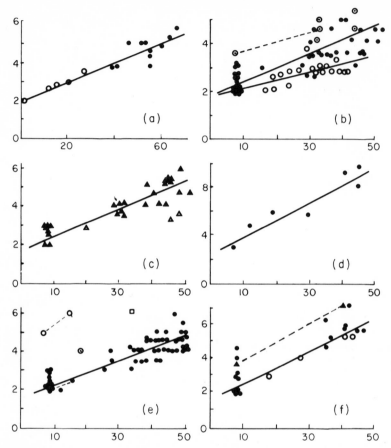

Figure 4.5. Clutch size (ordinate) in relation to latitude (abscissa). (a) The genus *Emberiza*. ○, African species nesting south of the equator; ●, species nesting in Europe and Asia. The two sets of points fit the same line. (b) The family Tyrannidae. ○, South American species; ●, North American species. The two sets of points fit lines of significantly different slope. ⊙, genus *Myiarchus* (hole nesters); ⊙---⊙, *M. tuberculifer*. (c) The family Icteridea. △, South American species; ▲, North American species. (d) The genus *Oxyura* (family Anatidae), worldwide distribution. (e) The Thraupinae (family Emberizidae) plus Parulidae, Central and North America. ● - - - ●, *Myioborus miniatus*; ○ - - - ○, *Euphonia lauta* (hole nester); □, *Protonotaria citrea*, hole nester; ⊙, *Euphonia luteicapilla*, niche nester. (f) The family Troglodytidae. ○, South American species; ●, North American species; ▲ - - - ▲, *Troglodytes aedon*. (From Cody, 1966.)

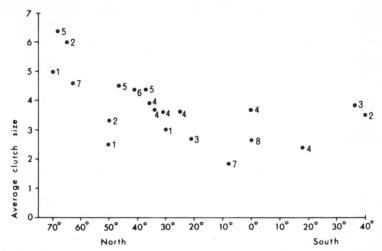

Figure 4.6. Latitudinal gradient of clutch size in owls. The numerals refer to the number of species used in calculating the average clutch size. (From von Haartman, 1971.)

ern latitudes, the Raven (*Corvus corax*) and the Red Crossbill (*Loxia curvirostra*) (Lack, 1947a).

Variations in clutch or litter size among other groups of animals are less well known. Increasing litter size with increasing latitude has been reported in nonhibernating prey mammals, but there is no trend in hibernating prey species (i.e., *Geomys*, *Thomomys*, and *Spermophilus*) or in predators (Lord, 1960). The clutch size of the slider turtle (*Pseudemys scripta*) is smaller in Louisiana than in Tennessee and Illinois (Cagle, 1950), and southern populations of the musk turtle (*Sternothaerus odoratus*) have smaller clutches than do northern populations (Tinkle, 1961), but the clutch size of the painted turtle (*Chrysemys picta*) is smaller in northern Michigan than in populations from Tennessee and Illinois (Cagle, 1954). There is no significant difference between the mean clutch sizes of tropical and temperate species of lizards, but tropical species usually lay several clutches, and temperate species usually lay a single clutch (Tinkle *et al.*, 1970). A reverse latitudinal trend in clutch size occurs in blackflies (Downes, 1964) and in marine bottom invertebrates (Thorson, 1950; Mileikovsky, 1971). Arctic blackflies lay a single clutch of from 20 eggs in the parthenogenetic *Gymnopais* to 150 eggs. Blackflies in southern Canada, however, lay from 200 to 500 eggs in the first gonotrophic cycle, and there can be two or more cycles each summer. The fewer eggs in the arctic is associated with larger size of the eggs, but unfortunately no data are available on survivorship of eggs in either area. The great fecundity of marine bottom invertebrates is well known, but it is limited to those species with a long period of planktotrophic pelagic larval life. Species that have a short planktotrophic pelagic larval life or lecithotrophic larvae or that are viviparous are characterized by a much lower fecundity. Predation is the most serious cause of mortality in marine bottom invertebrates, and the planktonic stage is most at risk. A long period of planktonic life results in low probabilities of reaching maturity (Thorson, 1950), raising the replacement rate and requiring a greater fecundity compared with the other groups.

Lack (1947a, 1954, 1968) suggested that clutch size tended to increase from western to central Europe, as especially shown by the European Robin (*Erithacus rubecula*) (Fig. 4.7). This seems to be part of a more general trend of smaller clutches in milder, maritime climates compared with clutch sizes in more severe, mainland and continental climates (Lack, 1947a, 1968; Cody, 1966, 1971). A smaller clutch in island forms compared with mainland forms has been reported in birds from Corsica and Sardinia (Lack, 1947a) and other birds (Fig. 4.8), in two species of salamanders

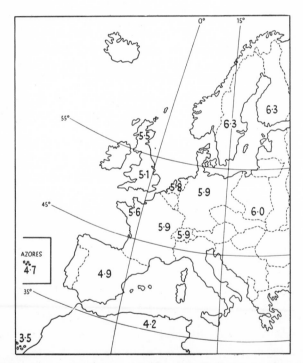

Figure 4.7. Average clutch sizes of the European Robin in several countries, showing increasing clutch size from south to north and from west to east. (From Lack, 1954.)

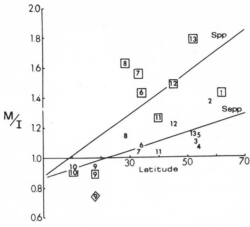

Figure 4.8. The ratio of mainland clutch size to island clutch size, M/I, in relation to latitude for various island groups. 1. Alaska/St. Lawrence Island. 2. Scotland/St. Kilda. 3. Continental Europe/England. 4. England/Ireland. 5. Continental Europe/Ireland. 6. California/Channel Islands. 7. North Africa/Madeira. 8. North Africa/Canary Islands. 9. Guatemala/West Indies. 10. Venezuela and Costa Rica/Trinidad. 11. Australia/New Zealand. 12. New Zealand/New Zealand islands. 13. Southern Chile/Falkland Islands. Boxed numbers, comparisons for endemic island species; regression line Spp: $Y = 0.884 + 0.014X$. Nonboxed numbers, comparisons for endemic subspecies; regression line Sspp: $Y = 0.864 + 0.006X$. The regression lines are significantly different at the 10% level. Number enclosed in diamond, comparison of endemic genera. (From Cody, 1971.)

(*Batrachoseps attenuatus* and *Aneides lugubris*) in the San Francisco Bay region (Anderson, 1960), in lizards (*Lacerta* spp.) in the Mediterranean Sea (Kramer, 1946), and in the cottonmouth (*Agkistrodon piscivorus*) from the Cedar Keys and mainland Florida (Wharton, 1966). In tropical species of birds, clutch size does not differ between island and mainland populations (Fig. 4.8).

Data on the variations in clutch size with elevation show no distinct trend. In the Song Sparrow in western North America, the largest clutch size occurs at high elevations in the Sierra Nevada range (Johnston, 1954). In Great Britain, the clutch size of the Meadow Pipit (*Anthus pratensis*) decreases with increasing elevation (ranging only from sea level to just over 1000 feet) (Coulson, 1956). Clutch size also decreases with elevation in several corvids in Great Britain (Holyoak, 1967) and in some Asian passerine species (Dementiev and Stépanyan, 1965). Clutch size in the salamander *Desmognathus ochrophaeus* is greater at higher elevations on Mount Mitchell in North Carolina, but this is associated with greater body size and later age of first reproduction at higher elevations (Tilley, 1973). In the mouse *Peromyscus maniculatus*, litter size increases and the number of litters per year decreases with elevation (Dunmire, 1960; Spencer and Steinhoff, 1968), but this is not the case in the bank vole (*Clethrionomys glareolus*) in Switzerland (Claude, 1970) and the pika (*Ochotona princeps*) in western North America (A. T. Smith, 1978). The bank vole has a single, small litter at 1700 meters. At 400 meters the voles have two larger litters, one in the spring and another in the fall. Pika populations are characterized by two litters per year throughout the species' range. Individuals at high elevations have smaller litters than do low-elevation individuals. In the Meadow Pipit, salamander, vole, and pika, mortality is greater at lower than at higher elevations.

Geographic variation in the physical environment provides only a first approximation of the length of the period favorable for breeding. Local conditions also play a role and will result in divergences from the overall trend. Local conditions include the effects of rainfall, availability of food, and presence of competitors and predators. An example of local effects on life history patterns is provided by the Great Tit, which in southern England breeds in both broad-leaved woods and pinewoods (Perrins, 1965; Lack, 1966; Krebs, 1970). The birds of the broad-leaved woods usually lay a single clutch, whereas as many as 28% of the birds of the Scots pine woods lays two clutches (Table 4.4). The factor responsible for this differ-

TABLE 4.4

Clutch Size and Success in Great Tits in England[a]

	First clutch		
Habitat	Mean clutch	Nestling weight	Percentage nestlings dying (excluding predation)
Broad-leaved woods	9.8	18.9	5
Scots pine woods	10.0	14.6	38
Corsican pine woods	9.1	14.0	40

	Second clutch		
Habitat	Percentage pairs starting second	Mean clutch	Nestling weight	Percentage nestlings dying (excluding predation)
---	---	---	---	---
Broad-leaved woods	2	7.3	15.9	41
Scots pine woods	28	7.6	17.4	13
Corsican pine woods	11	8.4	18.1	22

[a] From Lack (1966), Table 10.

ence is the availability of caterpillars (Lack, 1966). The caterpillar population in the oak woods drops off sharply following the first brood, but in the pinewoods it increases following the first brood. Great Tits in pinewoods have a better opportunity than those in oak woods to lay two clutches because of the availability of caterpillars for the second brood. Without information on the costs in adult survival, however, interpreting the differences in clutch size in the different habitats is not possible. For instance, because the oak woods' Great Tits are limited to a single brood, a larger clutch size in oak woods than in pinewoods might be expected from the assumptions of the model, but this expectation is contrary to observation (Table 4.4). However, the high fledging success and high weights of fledglings may result in a greater $\Sigma \lambda_x$ in oak woods than in pinewoods' populations, leading to a lower replacement clutch. Furthermore, the ease of gathering food in the oak woods may result in greater adult survivorship, again increasing $\Sigma \lambda_x$ and lowering the size of the replacement clutch, compared with pinewoods' populations.

Without detailed comparative data on λ_x patterns in various populations, much less for each genotype, we must be satisfied with general trends, at least until better data are available. We must search for explanations of the data we have rather than of the data we would like to have. Data divergent from the general trends indicate areas for further research.

Annual and Seasonal Variations

Environmental variability complicates the picture. A particular clutch size will have different effects on the probabilities of further life depending on conditions. Laying the average replacement clutch will depress survivorship when conditions are poorer than average and will increase survivorship when conditions are better than average (Fig. 4.2). Even without the burden of raising a brood, survivorship is depressed when conditions are below average and increased when conditions are better. Thus, when conditions are poor survivorship is reduced, raising the replacement rate and demanding an *added* reproductive effort, regardless of the cost. When conditions are better than average, survivorship improves, lowering the replacement rate and allowing a reduced reproductive effort just at the time when an extra reproductive effort can be obtained at no extra cost (Fig. 4.2). Attempting to maintain a constant clutch size in the face of varying environmental conditions seems an inefficient and untenable life history pattern.

If life expectancy is short relative to the period of environmental fluctuation, the population will not survive severe conditions because the low survivorship demands an increased fecundity, which depresses survivorship even further. These populations evolve dormant stages (e.g., spores, eggs, seeds) that survive the severe period. If life expectancy spans several periods favorable for breeding, but favorability is variable, we might expect that those individuals that reduce clutch size when conditions are poor and increase clutch size when conditions are better than average, with no cost in survivorship (horizontal arrows in Fig. 4.2), would be selected over individuals that sustain a constant clutch size regardless of conditions (vertical arrows in Fig. 4.2), if such behavior has a higher probability of being more successful in maximizing $\Sigma \lambda_x \mu_x$. Adjusting clutch size to prevailing conditions results in a stable adult survivorship and in increasing the probability of successfully raising the young. Attempting to maintain a constant clutch size results in fluctuating adult survivorship and fluctuating success in raising the young. The former life history seems more likely to be selected.

That clutch size varies from year to year depending on food availability is well known (Lack, 1947a, 1954, 1966, 1968; Klomp, 1970). In some years when voles are plentiful, the Short-eared Owl (*Asio flammeus*) lays up to nine eggs, but in other years it lays only four or five eggs (Lack, 1954). Few Tawny Owls (*Strix aluco*) lay any eggs when vole populations are low in southern England (Southern, 1970), and Short-eared Owls, Snowy Owls (*Nyctea scandiaca*), and Pomarine Jaegers (*Stercorarius pomarinus*) do not breed during lemming population lows at Point Barrow, Alaska (Pitelka *et al.*, 1955). Among passerines the Thick-billed Nutcracker (*Nucifraga caryocatactes*) lays four-egg clutches after a good hazelnut crop but only three-egg clutches after below-average hazelnut crops in Sweden (Lack, 1954).

Seasonal variation in the size of clutches in multibrooded species is common (Lack, 1947a, 1954, 1966; Klomp, 1970). In some species, clutch size increases during the breeding season and then declines (Fig. 4.9), and in other species the clutch size is greatest at the beginning of the breeding season (Fig. 4.10). Lack (1954, 1966) considered these changes to be the result of seasonal variations in food supply. The first clutches of the first group are laid earlier in the season than the first clutches of the second group (cf. Fig. 4.9 and 4.10), and thus they are laid before the food supply reaches its seasonal peak, whereas the breeding seasons of the second group do not begin until later, when food is at its peak. After the seasonal peak in food supply, the clutch sizes of both groups decline.

No doubt, selection results in timing the breeding season to coincide with the peak availability of food, as Lack suggested, but it does not follow that birds are raising as many young as their food supply allows. Even if parents are raising as few young as possible consistent with replacement, the breeding season should still coincide with peak availability of food because survivorship of both young and adults is enhanced compared with

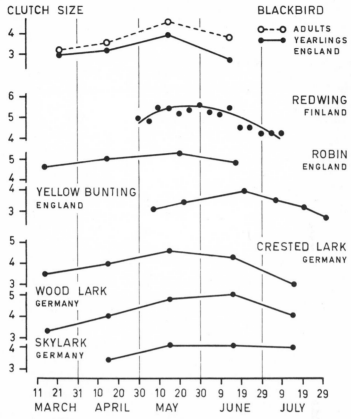

Figure 4.9. Seasonal variation in the clutch sizes of several European species. In these species the average clutch size rises to a peak before declining during the breeding season. (From Klomp, 1970.)

breeding at other times. Annual and seasonal variations in clutch size are not inconsistent with the hypothesis that parents raise as few young as they can and still replace themselves under prevailing conditions.

Variation with Body Size

The four assumptions from which life history patterns are to be predicted provide no expected relationship between clutch size and body size, but body size does affect life history parameters. Bonner (1965) showed that generation time lengthens as body size increases (Fig. 4.11). Larger animals live longer and require longer to mature than smaller animals, resulting in later ages of first and last reproduction and, therefore, longer generation times. Because populations are, on average, maintaining a

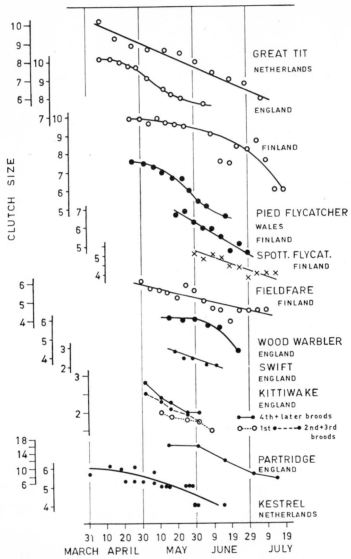

Figure 4.10. Seasonal variation in the clutch sizes of several European species. In these species the average clutch size declines throughout the breeding season. Note that these species begin breeding at a later date than those shown in Fig. 4.9. (From Klomp, 1970.)

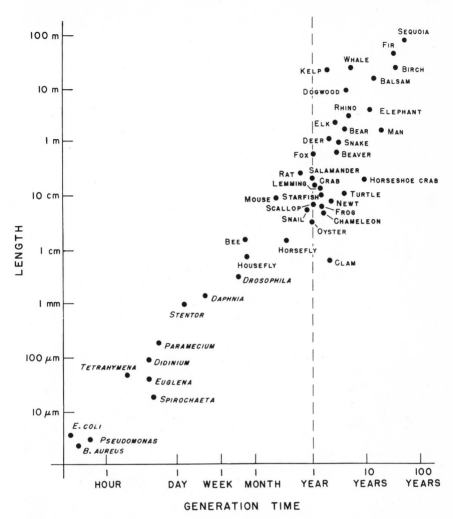

Figure 4.11. Relationship between body length and generation time of a variety of animals and plants. (From Bonner, 1965.)

"steady-state" size, individuals are, on average, just replacing themselves. We should expect, then, from Eq. (2.13), the population's rate of growth r to decrease as generation time increases. This is, in fact, the case (Fig. 4.12).

The size range of organisms in Figs. 4.11 and 4.12 is extraordinary. The small size of protozoans precludes great fecundity. They reproduce by fission and achieve high r by their very short generation times, less than 1 day. Among the larger organisms, animals such as elephants and whales

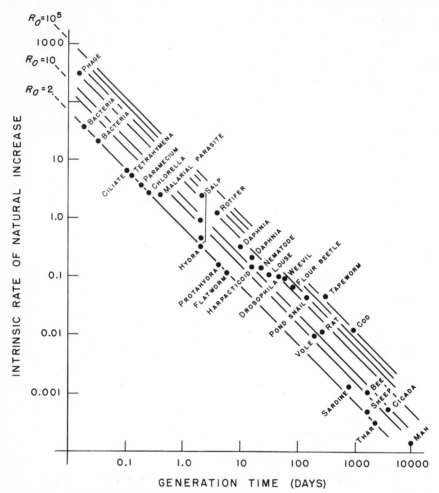

Figure 4.12. Relationship between generation time and the intrinsic rate of natural increase. (From Heron, 1972; after Smith, 1954.)

tend to have low fecundity, but trees such as sequoias produce millions of seeds. In the intermediate-sized classes, as well, individuals of different, similarly sized species can differ greatly in their fecundity. Body size does not seem to be a good predictor of fecundity.

Among birds, large albatrosses and small storm-petrels have single-egg clutches, and clutches of over 15 eggs occur in the small Blue Tit (*Parus caerulea*) and the large Ostrich (*Struthio camelus*). In both cases, the smaller species matures at an earlier age. Clutch size, however, is not a good measure of annual fecundity because small birds that lay several clutches per season may lay as many eggs as larger birds do with a single clutch. Prairie Warblers (*Dendroica discolor*) in Indiana lay, on average, 11 eggs in three clutches each year (Nolan, 1978). The clutch size is smaller but annual production is greater than in many birds laying a single clutch, as in ducks (Lack, 1968). Comparison of clutch sizes without considering the effects of age of first reproduction, number of clutches laid per year, or average life expectancy can provide only misleading information regarding either the dynamics of population growth or the evolution of clutch size.

Among lizards, however, Tinkle *et al.* (1970) reported increasing clutch size with body size; the smaller, earlier-maturing (12 months or less), multiple-brooded species lay smaller clutches than larger, later-maturing (24 months or more), single-brooded species (Fig. 4.13). Body size, however, is not the factor determining a species' life history pattern. Species of the same size can be either early-maturing or late-maturing or either single-brooded or multiple-brooded.

Small, early-maturing, multiple-brooded lizards tend to be tropical and short-lived (Tinkle, 1969; Tinkle *et al.*, 1970). The short life expectancy selects for early maturity and high fecundity, but the longer breeding season in the tropics allows the eggs to be laid in a series of small clutches. In the longer-lived temperate species, the large clutch sizes result from the combination of late maturity and single clutch. There are exceptions, and, to evaluate these, data on survivorship, the factors affecting successful reproduction and the length of the breeding season will have to be obtained. Evidently, however, body size is not the factor determining clutch size in lizards.

On intraspecific comparison, however, clutch size does increase with increasing body size in species with indeterminate growth, as in fishes (Williams, 1966a), salamanders (Tilley, 1973), and lizards (Tinkle *et al.*, 1970). Such a phenomenon is easy to understand. The number of young that can be produced is a function of size, and, as individuals grow larger after they have matured, the number of eggs or young can increase with age. An attempt to maintain a constant clutch size at every age would place great energy demands on the smaller, younger animals with high costs in survivorship. It is no surprise that clutch size increases with size in species with indeterminate growth.

Clutch Size and Primary Productivity

Although most research on the clutch sizes of birds has concentrated on explaining differences between them, Ricklefs (1970) drew attention to the striking constancy of clutch size when considering the diversity of feeding behavior of the species involved, from fly-catching to ground-scratching, and the 10-fold range of primary productivity rates between the habitats in

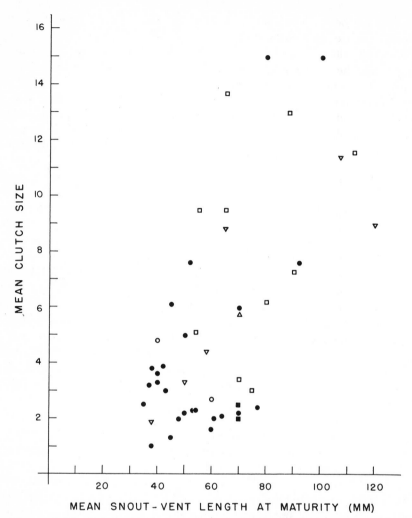

Figure 4.13. Relationship between the body length at maturity and clutch size in lizards. Single-brooded species are represented by open symbols; multiple-brooded species, by closed symbols. The age of maturity is given by the following symbols: ○, 12 months or less; △, 18 months; □, 24 months; ▽, 30 months or more. (From Tinkle *et al.*, 1970.)

which birds breed. As most theories (reviewed below) assume some relationship between food availability and clutch size, the observed lack of correlation between clutch size and foraging habits and, especially, primary productivity seems contrary to those theories. Why individuals living in a habitat of low productivity should be able to find as much food for providing for their young as individuals living in habitats of high productivity has not been explained. Why a habitat can supply enough food to maintain a denser population of a species than a less productive habitat but not enough to allow the evolution of a larger clutch there is puzzling, if one believes that the availability of food determines not only the "steady-state" population size but the clutch size.

As suggested earlier, in the basic model of food-limited populations (Fig. 3.5), the amount of food in a habitat does not affect survivorship until the UCD is reached, above which declining per capita consumption has the same effect on survivorship, regardless of population size. It seems more probable that survivorship would vary with environmental quality, which includes the availability of food, declining as conditions become more severe (Fig. 3.19). Nevertheless, as survivorship declines, the replacement birth rate increases (Table 2.6), requiring either a shorter generation time or greater fecundity, or both, if individuals are to replace themselves or populations are to maintain a "steady state" (Fig. 3.12). However, as noted before, as conditions worsen, it is more likely that breeding will be postponed to a later age or fecundity will be reduced, or both. Birth and death rates simply cannot vary to the same extent as primary productivity can vary. It appears that any habitat must supply the minimum resources necessary to ensure that an average individual can live long enough to produce the young that will replace itself and its mate. Abundant resources do not result in individuals increasing their clutch size, if clutch size evolves as suggested above rather than as suggested by Lack. Instead, abundant resources result in a higher UCD and larger population size (Fig. 3.5).

Data on survivorship in natural populations are scarce. In the best studied group of animals, the birds, nesting mortality is relatively uniform for open-nesting species and for hole-nesting species (Nice, 1957; Ricklefs, 1969), but a little higher for each group in the tropics (Skutch, 1966; Ricklefs, 1973). Adult annual mortality rates are remarkably uniform for birds of similar size (Hickey, 1952; Farner, 1955), and reviews of avian life history information often arrange data by taxonomic group rather than by primary productivity of habitats (Lack, 1968; Cody, 1971; von Haartman, 1971; Ricklefs, 1973). Variations in the life history data are small compared with the 10-fold difference in primary productivity of the habitats in which birds live.

Clutch Size and Parental Care

Skutch (1949) argued that, if the size of the clutch is determined by the amount of food that can be delivered by the parents, in those species in which only one parent provides for the young the clutch size should be smaller than in those species in which both parents provide for the young.

He then presented a list of six pairs of species resembling each other in nidification, habitat, and diet. In one species of the pair, only one parent fed the young; in the other, both parents did. Yet the clutch sizes did not differ. Skutch concluded that clutch size in these species was not determined by the amount of food the parents could deliver to their young. Additional examples have been provided by von Haartman (1955) and Skutch (1976). In many species of birds, parents are assisted in raising their young by their older offspring (Skutch, 1961, 1976; Fry, 1977). Not all helpers at the nest provide food for the young, but some do, and there is no indication that clutch size is increased in these species.

If clutch size has evolved as I suggest in this chapter, a larger clutch size in species in which both parents provide food to the young is not expected. Indeed, if the male's assistance reduces the female's effort, raising survivorship and lowering the size of the replacement clutch, the clutch size of species in which males help their mates and of species with other helpers at the nest should be smaller than in those species in which the female raises the young alone. Whether the clutch size in fact is increased depends on how much the replacement clutch is raised when the male does not help. Returning to the numerical example (Table 4.2), if the lack of paternal or other helper care raises the replacement clutch from 5.333 to 6.0 eggs, the clutch size will not change at all.

Age of First Reproduction

An important life history parameter is the age of first reproduction. Early reproduction increases ρ, provided that the survivorship schedule does not change (Cole, 1954). Thus, the postponement of breeding comes about because early breeding reduces λ_x values, by lowering survivorship either of the early breeder or of its offspring relative to λ_x values of late breeders or their offspring. The early breeder may lack the stamina or skill for hunting for suitable food in sufficient amounts to ensure its own and its offsprings' nourishment. Sometimes, however, breeding can be delayed, especially in males, by territorial behavior of older, more experienced individuals, such as occurs in the Red-winged Blackbird (*Agelaius phoeniceus*) (Orians, 1961). The postponement of breeding, however, has a cost. Every year that reproduction is postponed results in an ever increasing minimum clutch size (Table 4.2). Thus, it is not surprising that delayed reproduction occurs primarily in long-lived species, such as seabirds and raptors (Lack, 1954, 1968; Ashmole, 1963, 1971; Amadon, 1964), elephants, whales, and human beings (Asdell, 1964).

Hole-nesting species of birds are characterized by larger clutch sizes, longer nestling periods, slower growth rates, and greater fledging success than are open-nesting species (Lack, 1947a, 1954, 1966, 1968; Nice, 1957; Skutch, 1966, 1976; Ricklefs, 1969). The high survivorship of eggs and young combined with the large clutch size of hole-nesting species compared with open-nesting species appears to contradict the prediction of the four assumptions that large clutch size is a consequence of low survivorship. The explanation was anticipated partially by Skutch (1976). Populations of hole-nesting species are often limited by the number of available holes, as indicated by striking increases in population size, at times as much as 10-fold, when nest boxes are provided (Lack, 1954; von Haartman, 1971). Skutch (1976) suggested that competition for nest sites prevents many individuals from breeding and that production of a larger brood would make up for delays in the breeders' establishing themselves in suitable cavities and for those members of the species that fail to acquire a hole at all [contrary to Fretwell (1969), who hypothesized that competition for nest sites should result in selection for smaller broods]. Stated in terms of the model presented here, competition for nest sites prevents many individuals from breeding, effectively postponing the age of first reproduction and reducing the average clutch size, and thus raises the minimum replacement clutch (Table 4.2).

In an analysis of clutch size data on the hole-nesting Eastern Bluebird (*Sialis sialia*), Peakall (1970) showed that the largest clutch sizes occurred in the central portion of the species' range in eastern North America, where the breeding season is longest and where the populations are densest. The long breeding season again suggests a small clutch size, but intraspecific competition for holes should be most severe where population density is greatest. Greater competition results in a lower probability of acquiring a hole, increasing the average age of first reproduction and therefore increasing the replacement clutch size.

The Evolution of Semelparity

Sometimes the age of first reproduction is the age of last reproduction, such as occurs in periodical cicadas, mayflies, and Pacific salmon. Those individuals that produce a single clutch during their lifetime are called semelparous (Cole, 1954). As indicated earlier (p. 80). Cole (1954) argued that the evolution of an additional egg to the clutch of a semelparous individual of an annual species is equivalent to the evolution of multiple clutches (iteroparity) and immortal life. He assumed that the addition of a single egg to the clutch would be more likely than the evolution of a series of adaptations that would allow the organism to survive several periods of dormancy, and thus he asked why iteroparity occurred at all.

Although Cole (1954) phrased the question in terms of annual versus perennial species, some semelparous species are long-lived and some iteroparous species are short-lived. In my view, whether individuals are semelparous or iteroparous depends on the age of first reproduction and the expectation of life following reproduction. If the age of first reproduction is postponed to near the end of life, the individuals will be semelparous. Sometimes the time of first reproduction is not postponed by selection against early reproduction per se but is imposed on individuals by environmental factors. For example, the life history of anadromous fishes includes a migration between their freshwater spawning grounds and the ocean, where development to adulthood occurs. The age of first reproduction is determined by (1) the replacement clutch size, which increases with each passing year, and (2) the number of eggs the female can produce, which increases with each passing year, but the replacement rate is increasing more rapidly than the potential reproduction (Fig. 4.14). The female must breed before the replacement rate surpasses its ability to

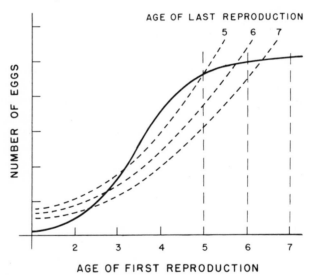

Figure 4.14. Replacement clutch size and potential clutch size of females in several hypothetical situations. Probabilities of survivorship (λ_x) and pattern of growth in body size are approximately the same in all individuals. The replacement clutch size, represented by the dashed lines, is a function of the age of first reproduction and the age of last reproduction. It increases rapidly as the age of first reproduction is postponed but is always lower for longer-lived individuals for any given age of first reproduction. The potential number of eggs, represented by the solid line, is a function of the size of the female and increases with age. The potential number of eggs increases rapidly at first, and then more slowly. In this set of examples, only the short-lived individuals can be semelparous because only they are able to lay a replacement clutch at their age of last reproduction. Whether individuals are semelparous or iteroparous depends on the factors that determine the ages of first and last reproduction.

produce a replacement number of eggs. Where rivers are long and swift only the largest fishes have the strength and stamina to reach the spawning grounds with sufficient energy to spare for spawning, and thus spawning occurs late in life, and only once, as is the case in Pacific salmon (*Onchorhynchus* spp.). In northeastern North America, where rivers are short and less swift, local populations of the Atlantic salmon (*Salmo salar*) breed at an early age and are iteroparous (Schaffer and Elson, 1975). Furthermore, among populations of the Atlantic salmon, the mean age of first breeding increases with the difficulty of upstream migration.

Summary

An animal's life expectancy is a consequence of the inevitable probabilities of dying from predators, disease, starvation, and accident, all of which are raised by reproductive activity. Selection always tends to extend life expectancy by reducing the effects of predators and disease organisms, increasing ability to find, capture, and transport food, and reducing the costs of reproduction. Nevertheless, a minimum amount of reproduction is required if individuals are to leave offspring in the next generation. The minimum number of eggs laid or young born depends not only on life expectancy but on the length of the breeding season, the development time from birth to independence, and the first and last ages of reproduction. The age of first reproduction is determined by the individual's physiological development and environmental quality, and it will occur as early as successful reproduction becomes probable. Whether a species is single- or multiple-brooded depends on the relative lengths of development time and breeding season, a short development time and long breeding season leading to multiple broods. The availability of food determines primarily the time of breeding and secondarily the clutch size. Food availability does not determine the maximum number of eggs that can be laid or young provided for but rather the number that can be raised at the minimum cost in survivorship and still replace both the female and her mate. Semelparity is a consequence of the age of first reproduction being postponed until the latest age at which breeding can be undertaken successfully.

PREVIOUS THEORIES

Clutch Size

The great diversity of clutch size variations has stimulated several theories. These can be briefly categorized as "maximal" reproduction, "adjusted" reproduction, and "optimal" reproduction.

The best known, most widely accepted theory was first proposed by David Lack in 1947. It has been frequently restated in different forms (Lack, 1948a, 1954, 1966, 1968), perhaps the most representative being (Lack, 1968), "The number of eggs in the clutch has been evolved to correspond with that from which, on the average, the most young are raised. In nidicolous species, the limit is set by the amount of food which the parents can bring for their young, and in nidifugous species by the average amount of food available for the laying female, modified by the size of the egg" (p. 5). The hypothesis was quickly applied to other groups of animals (Lack, 1948b, 1954).

Most of the criticism of Lack's hypothesis has centered on the relationship between clutch size and the availability of food, and several authors have drawn attention to factors that limit the amount of energy that parents can deliver to their young. Cody (1966) suggested that clutch size could be reduced and that the smaller clutch size would be selected if the conserved energy were allocated to (1) increasing the probability of escaping predators, (2) improving the ability to compete, or (3) providing more energy for each young. Brockelman (1975) has extended the discussion on the effect of competition on clutch size. Royama (1969) considered the effects on clutch size of the energy demands of the young, and Smith and Fretwell (1974) considered the trade-off between clutch size and size of the young. In my opinion, none of these discussions challenges Lack's (1954) hypothesis that clutch size is determined by the amount of energy available, considering the limitations set by such factors as the "food requirements of the young, . . . the interval at which the eggs of the brood hatch, the rate of growth of the nestlings, the predation-rate, and the part, if any, played by each parent in raising the young" (pp. 43–44).

According to Lack's hypothesis, the most frequent clutch size should be the most productive. The results of published work, reviewed by Klomp (1970), are mixed. In the Common Swift (*Apus apus*), one of the earliest species studied, the most frequent clutch in England was two eggs, but three-egg clutches were common and four-egg clutches extremely rare (Lack and Lack, 1951; Lack, 1954). In food-poor years two-egg clutches were most productive, and in food-rich years three-egg clutches were most productive. Later, when an additional nestling was added to nests of three young (Perrins, 1964), the parents of four raised fewer young than did parents of three young (Table 4.5). These results indicate that four-egg clutches are rare in England because they are less productive than three-egg clutches and that two-egg clutches are frequent because they are most productive in food-poor years.

Among the passerine species studied, however, fledging success is usually independent of clutch size, the largest clutches producing the most

TABLE 4.5

Reproductive Success of Experimental Broods in the Swift (*Apus apus*)[a]

Brood size	Number broods	Number young	Number lost	Percentage lost	Percentage fledged per brood
1	30	30	2	6.7	0.933
2	72	144	5	3.5	1.931
3	20	60	9	15.0	2.550
4	16	64	36	56.3	1.750

[a] From Perrins (1964).

fledglings. Lack (1954, 1966, 1968) argued that the largest broods were not necessarily the most successful if the young from the larger broods left the nest underweight and if this reduced the probability of surviving compared with heavier fledglings from smaller broods. It has been shown that nestlings of large broods often fledge at below average weight, as in the Starling in Europe (Lack, 1948c) and the United States (Crossner, 1977), the Great Tit in Great Britain (Lack *et al.*, 1957; Perrins, 1965), and the Pied Flycatcher (*Muscicapa hypoleuca*) in Finland (von Haartman, 1955; Tompa, 1967), as well as in young from experimentally enlarged broods (see below).

In order to determine postfledging survival, the recovery rates of banded birds from small and from large broods were compared, either of birds at least 3 months after fledging or of birds returning in the following breeding season. The young of large broods do not survive as well as those from small broods in the Starlings in Switzerland, England, and the Netherlands (Lack, 1948c), in the Great Tit in England, although the most productive clutch size varies from year to year depending on the food supply (Lack *et al.*, 1957; Lack, 1958, 1966; Perrins, 1965), and in the Pied Flycatcher in Finland (von Haartman, 1955, 1967). In other populations, however, postfledging survival was independent of brood size, as in the Pied Flycatcher in Germany (Curio, 1958) and Great Britain (Lack, 1966) and in the Song Thrush (*Turdus ericetorum*) in Great Britain (Lack, 1949). Sometimes when postfledging survival is lower in the larger broods, the difference in survivorship is not sufficient to offset the larger initial size of the brood, resulting in greater productivity of the larger but less frequent brood size, as in the European Blackbird (*Turdus merula*) in Great Britain (Lack, 1949).

Klomp (1970) thoroughly reviewed the literature, most of which includes papers of less detail or completeness than those already mentioned. A persistent problem throughout this kind of research is whether the observed variations in clutch size are owing to genetic differences among individuals or are individual responses to environmental differences. In the most studied species, the Great Tit, clutch size varies with the available food supply, the date on which the clutch is started, the age of the female, the density of the breeding pairs, and even the section of the woods inhabited (Perrins, 1965; Lack, 1966; Krebs, 1970). Allowing for this variation, Perrins and Jones (1974) were able to show that the total heritability in clutch size in the Great Tit was 0.51. This does not mean that a particular size is inherited. Instead, females inherit a tendency to lay a particular clutch size in relation to the average size of the total population. For example, a female inherits a tendency to lay a larger than average clutch. Her clutch size will vary from year to year as the population's average clutch size varies from year to year in response to environmental conditions, but her clutch will always be larger than average under whatever conditions prevail.

The fact that clutch size is inheritable does not ease the problem of how to interpret observable clutch size variations, inasmuch as an individual with a given genotype can produce a range of clutch sizes depending on conditions. Furthermore, Mountford (1968) showed how this variability can result in the most frequent clutch size being smaller than the most productive. He assumed (1) a normal frequency distribution of clutch sizes for a given genotype, (2) a declining proportion of survival of young from broods above some minimal clutch size, and (3) the productivity of a genotype is the total number of offspring W from all clutches N of individuals with the same genotype. Then,

$$W = N \sum_x xf(x)p(x),$$

where $f(x)$ is the frequency of clutch size x, and $p(x)$ is the proportion of young fledging from clutch size x. Mountford's numerical example is given in Table 4.6, and it shows the most frequent clutch size to be five and the most productive to be seven.

The most productive clutch size is not necessarily the most frequent because the selected genotype is the one that produces the most offspring from the whole range of clutch sizes characteristic of the genotype. If each genotype has the same frequency distribution $f(x)$ around a different modal clutch, the number of offspring fledging from the clutches of each genotype can be calculated. In Mountford's numerical example, the most productive genotype has a modal clutch of five, and this genotype will be selected rather than the genotype whose modal clutch is seven (Table 4.7). Whether the most frequent clutch is greater than, less than, or the same as the most

TABLE 4.6

Frequency and Success of Clutch Sizes for a Given Genotype[a]

Clutch size x	Relative frequency $f(x)$	Proportion of young fledged $p(x)$	Average number fledged per clutch $xp(x)$
1	0	1	1
2	0	1	2
3	0.08	1	3
4	0.16	0.99	3.96
5	0.21	0.975	4.88
6	0.19	0.92	5.52
7	0.15	0.82	5.74
8	0.11	0.67	5.36
9	0.07	0.44	3.96
10	0.03	0	0
11	0	0	0

[a] From Mountford (1968).

productive clutch depends on the frequency distribution of clutch sizes and the proportion of young surviving from each clutch size.

If Mountford's view bears any relation to reality, then in a genetically diverse population, clutches of the same size could in fact represent different genotypes, and clutches of different sizes could represent the same genotype, making it virtually impossible to compare the successes of clutches of different sizes as a test of Lack's hypothesis. Klomp (1970), however, believes that differences in the clutch sizes exhibited by a

TABLE 4.7

Numbers of Fledged Offspring from Clutch Size Frequency Distribution Centered on Different Modal Clutch Size (μ)[a]

μ	Number of fledged offspring	μ	Number of fledged offspring
3	3.68	8	3.24
4	4.33	9	2.60
5	4.66	10	1.09
6	4.45	11	0.32
7	4.09	12	0.00

[a] From Mountford (1968).

genotype are adaptive modifications to environmental conditions and that when clutch sizes are studied under as similar conditions as possible the observed variations can be considered hereditary. According to Klomp (1970), in the several cases in which a discrepancy occurs between the most frequent and most productive clutch sizes, the investigator failed to exclude sufficiently adaptive modifications. In other words, data that show the same clutch size to be the most frequent and most productive, as expected from Lack's hypothesis, are acceptable, whereas data that show the most productive clutch size to be larger than the most frequent one are suspect.

Unfortunately, the information needed to evaluate adequately the success of a particular clutch size is (1) the survivorship of individuals laying clutches of different size, (2) the survivorship of the young from clutches of different size, not only through the first 3 months or year but throughout their lives, and (3) the reproductive success of the young coming from different clutch sizes. And this assumes that individuals with different genotypes can be distinguished. This seems an almost impossible task. Perhaps, as in other fields of scientific inquiry, we may have to test our ideas by erecting alternative models, which make axiomatic assumptions from which predictions are deduced, and by comparing the predictions of the alternative models with empirical evidence.

Lack's hypothesis has stimulated the so-called twinning experiments in which one or more eggs or young of appropriate age are added to the clutch or brood of a nesting bird. If Lack's hypothesis is correct, the parent bird(s) should be unable to raise successfully a larger than normal brood. "Twinning" experiments, then, should indicate high rates of egg or chick mortality or the fledging of underweight chicks. The results of "twinning" experiments are mixed (Table 4.8). Several species (e.g., Swallow-tailed Gull, North Atlantic Gannet, South Atlantic Gannet, Herring Gull on Lake Erie) seem capable of raising larger broods than normal without apparent detriment, but most cases of "twinning" indicate that parents cannot raise chicks of enlarged broods to normal weight. Losses, however, are not always attributable to starvation because they occur in the early days of nestling life at a time when the energy demand of the enlarged brood is less than for older nestlings of a normal brood. The causes of losses are usually unknown, but in the case of the Lesser Black-backed Gull and Herring Gull in England (Harris and Plumb, 1965) the parents were unable to protect the young of enlarged broods from inclement weather. Although only three of the seven young Herring Gulls that reached 200 gm weight were fledged, all 20 young Lesser Black-backed Gulls that reached 200 gm weight did. The average fledging weight of gulls from normal broods is about 700 gm, which indicates that food is not the limiting factor on clutch size in this species.

TABLE 4.8

Results of "Twinning" Experiments

Species	Normal clutch	Experimental manipulation	Sample nests	Results	Reference
Laysan Albatross (*Diomedea immutabilis*)	1	Add 1 egg Add 1 chick	18 18	Cannot hatch 2 eggs; only rarely rear 2 chicks; chicks underweight	Rice and Kenyon (1962)
Manx Shearwater (*Puffinus puffinus*)	1	Add 1 chick	9	1.56 young fledged/pair; chicks very underweight	Harris (1966); Perrins *et al.* (1973)
Manx Shearwater	1	Add 1 chick	42	1.79 young fledged/pair; lighter twin survived and returned as well as singles	Perrins *et al.* (1973)
Short-tailed Shearwater (*Puffinus tenuirostrus*)	1	Add 1 egg Add 1 chick	20 20	Poor hatching success; poor fledging success; chicks underweight	Norman and Gottsch (1969)
North Atlantic Gannet (*Sula bassana*)	1	Add 1 egg Add 1 chick	17 13	Parents can hatch 2 eggs and raise 2 chicks (of the same age), which are slightly underweight	Nelson (1964)
South Atlantic Gannet (*Sula capensis*)	1	Add 1 egg	54	Pairs can raise enlarged broods, but fledglings underweight; parents not underweight, and some returned to breed	Jarvis (1974)
White Booby (*Sula dactylatra*)	2	Add 1 chick to nests with 1 chick	16	Parents could raise two chicks for 2 weeks, but 1 chick always disappeared	Dorwood (1962)
Brown Booby (*Sula leucogaster*)	2	Add 1 chick to nests with 1 chick	8	Parents could raise 2 chicks for longer than usual before 1 died or disappeared	Dorwood (1962)

(continued)

TABLE 4.8. (*Continued.*)

Species	Normal clutch	Experimental manipulation	Sample nests	Results	Reference
Swallow-tailed Gull (*Creagrus furcatus*)	1	Add 1 egg Add 1 chick	31 30	No difference from controls in hatching, growth, and fledging success, and in return rates of adults to colony	Harris (1970b)
Glaucous-winged Gull (*Larus glaucescens*)	3	Add 1, 2, or 3 pipped eggs or chicks	97	Broods of 4, 5, or 6 chicks as successful as normal broods	Vermeer (1963)
Herring Gull (*Larus argentatus*) and Lesser Black-backed Gull (*Larus fuscus*) (England)	3	Add 1, 2, or 3 chicks	10	Unsuccessful—parents could not protect chicks from elements, but *fuscus* could feed enlarged broods	Harris and Plumb (1965)
	3	Add 1, 2, or 3 chicks	10		
Herring Gull (Canada)	3	Add 1 or 2 chicks, or subtract 2 chicks	50	Pairs successful in raising enlarged broods; no differences in fledging weights	Haymes and Morris (1977)
Razorbill (*Alca torda*)	1	Add 1 chick	12	Three pairs raised twins; most losses early; twins underweight when fledged	Plumb (1965)

Species	Clutch/brood size	Manipulation		Result	Reference
Razorbill	1	Add 1 chick	14	1.14 fledglings/nest; "twins" underweight	Lloyd (1977)
Atlantic Puffin (*Fratercula arctica*)	1	Add 1 chick	4	Two pairs raised "twins," but "twin" lighter	Corkhill (1973)
Common Swift (*Apus apus*)	2–3	Add 1 chick	16	Nestling mortality always greater in broods of 4	Perrins (1964)
White-bearded Manakin (*Manacus manacus*)	2	Add 1 chick	6	Unsuccessful; female apparently could not feed young efficiently	Lill (1974)
House Martin (*Delichon urbica*)	2–5 (first brood)	Subtract 1 chick from 2; add 1 to 5	12	Enlarged broods completely successful	Bryant (1975)
	1–4 (second brood)	Add 1 to 4	15		
House Finch (*Carpodacus mexicanus*)	2–3	Add 1 or 2 eggs	3+	Nestlings seemingly normal at fledging	Wagner (1957)
Yellow-eyed Junco (*Junco phaeonotus*)	2–3		2+		

In a more sophisticated "twinning" experiment with the Starling, Crossner (1977) not only manipulated brood sizes but supplied additional food just outside some of the nest boxes. In pairs without extra food, the average fledgling weight increased with brood size up to three chicks, dipped slightly through brood size 6, and dropped precipitously through brood size 9 (Fig. 4.15). No pairs fledged 10 chicks. Pairs supplied with *ad libitum* food were able to raise up to 10 chicks (the largest brood size attempted), and all chicks fledged at nearly the maximal weight attained under natural conditions (Figs. 4.15 and 4.16). Although the average clutch size of the Starling in New Jersey, where these experiments were conducted, was 4.68, lower than evidently could be raised by Starlings in this area, Crossner (1977) interpreted his results as being consistent with Lack's hypothesis.

Schifferli (1978) added a single chick to broods of three, four, and five at hatching of the House Sparrow (*Passer domesticus*) in England and compared the success of the parents of these artificially enlarged broods with the success of parents of natural broods of the same size, four, five, and six. In the control broods, the average weight of nestlings (age 10 days) declined with increasing clutch size. In the experimental broods, the average weight of nestlings at 10 days of age was not only below the average of control broods of the same size but below the average of the control brood of the clutch size the parents originally laid. Birds that laid

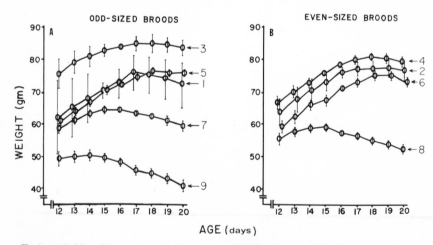

Figure 4.15. The average weights of nestling Starlings, which did not receive extra food, divided into odd- and even-sized broods. The numbers on the right indicate brood size. The vertical lines represent least significant intervals. Comparisons between brood sizes are appropriate only within age classes. Comparisons between ages are not appropriate. (From Crossner, 1977.)

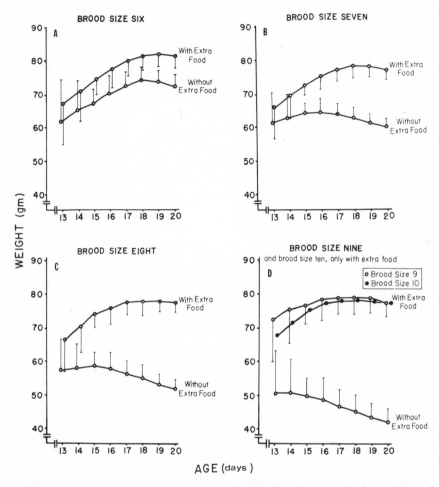

Figure 4.16. The average weights of nestling Starlings from broods of 6, 7, 8, 9, and 10, comparing broods receiving and not receiving additional food. The vertical lines represent food supply least significant intervals. The appropriate comparison is between food supply for a given clutch and at a given age. Between brood size and between age comparisons are statistically inappropriate. No Starlings raised broods of 10 without additional food. (From Crossner, 1977.)

three, four, or five eggs evidently could not raise successfully a brood from a larger clutch, giving further support to Lack's hypothesis.

Cody (1971) and Hussell (1972) criticized interpretations of "twinning" experiments in which the parents did not successfully rear the enlarged broods. Such results are usually taken as evidence supporting Lack's hypothesis, but, if a particular clutch size has evolved independently of the food supply, there is no reason to assume that parents should be able to raise enlarged broods because behavioral patterns as well as clutch size are evolving. If a clutch size of three has been selected, one should not expect, for example, food-gathering behavior to be selected to provide for clutches of four or five. Regardless of whether Lack's hypothesis is correct or incorrect as an explanation of the evolution of clutch size, we should expect that most populations will have evolved food-gathering behavior to be no more efficient than necessary to supply the chicks of a normal clutch.

Hussell (1972) argued further that in the alternative case, when parents were able to raise an enlarged brood, the experiments unequivocally showed that the food supply in relation to food-gathering ability was more than adequate to raise a larger brood. Such experiments, however, do not constitute a refutation of Lack's hypothesis if the environmental conditions of the experiment are different from those prevailing when clutch size and food-gathering behavior evolved, as has been argued by Nelson (1964) and Lack (1966) in the case of the North Atlantic Gannet. Seemingly, "twinning" experiments cannot provide unambiguous evidence in favor of one hypothesis or another.

Lack's hypothesis has proved to be heuristic, stimulating a great amount of research and speculation. The hypothesis has been widely applied by Lack, his colleagues, and his disciples to explain geographic, annual, and seasonal variations in the clutch sizes of many species of birds. "But, unfortunately, the wide applicability of the theory depends on the existence of largely undemonstrated food variations which must *ex hypothesi* follow observed clutch size trends," and, as Cody (1969) continued, "This can, at times stretch our credulity" (p. 1186). Furthermore, Lack's hypothesis has never satisfactorily accounted for those cases in which females who care for their young alone raise as many young as females of closely related species in similar situations who are assisted by their mates (Skutch, 1949, 1976; von Haartman, 1955, 1971). Another puzzle is the lack of correspondence between clutch size and the primary productivity of the various habitats in which a species lives (Ricklefs, 1970). Lack's hypothesis has not accounted for the increasing clutch size with latitude of nocturnal predators (Fig. 4.6) or nocturnal mammals (Lord, 1960), although Owen (1977) tried to save Lack's hypothesis by suggesting that greater prey diversity at low latitudes results in less efficient hunting and thus accounts for the smaller clutch size of not only tropical owls but other tropical species.

A second hypothesis, sometimes referred to as "adjusted" reproduction, was proposed by Skutch (1949, 1967, 1976) and Wynne-Edwards (1955, 1962). According to this hypothesis, clutch size evolves in response to a population's mortality rate. High mortality rates result in the evolution of large clutches, and low mortality rates result in small clutches. In particular, the small clutch sizes of long-lived tropical passerines and near-passerines and long-lived seabirds have evolved to prevent overpopulation and destruction of the species' environment. Skutch (1949) enlarged two broods of the Song Tanager (*Ramphocelus passerinii*) for 1 day each and found no difficulty for the parents to feed the young. Wagner (1957) enlarged several broods of House Finches (*Carpodacus mexicanus*) and Yellow-eyed Juncos (*Junco phaeonotus*), which successfully fledged. Both Skutch and Wynne-Edwards, however, explicitly introduced group selection to account for the evolution of a reduction in clutch size, and therefore this hypothesis has not received the attention it deserves.

A more recent development is the notion of "optimal" reproduction, which takes into consideration the costs of present reproduction in terms of future reproduction. An individual's lifetime reproduction can be increased by reducing the clutch size below the maximum that parents could raise successfully, if the reduction in present reproduction increases the probability of future reproduction. The idea has been analyzed in terms of the effect of present reproduction on residual reproductive value (Williams, 1966a, b), the survival rates of young and adults (Charnov and Krebs, 1974), and the compromise between individual progeny fitness and total parental fitness (Pianka, 1978). These analyses are theoretical arguments developed to show how the most frequent clutch size could be less than the most productive clutch size, an idea that Lack (1954) anticipated and later accepted (Lack, 1968) without modifying his main thesis. The notion that reproduction has a cost in survivorship is certainly correct, but it has not been applied to any of the many variations in clutch and litter sizes, and I believe that it cannot predict variations in clutch and litter sizes without additional assumptions, such as my assumption 4. With another elaborate model relating reproductive risk and adult mortality in birds, Ricklefs (1977) showed an inverse relationship between fecundity and longevity, a not unexpected result, but as with other models of optimal reproduction his model does not, and I believe cannot, predict the clutch sizes of birds.

Elliott (1975), apparently independently, applied the idea of optimal reproduction to the evolution of polygyny in yellow-bellied marmots (*Marmota flaviventris*). Females of monogamous pairs have larger litters than do females of polygynous groups, yet polygyny is more frequent than monogamy. Elliott suggested that lifetime production of females who joined harems was greater than that of females of monogamous pairs because the former lived longer as a result of a better chance of spotting and avoiding predators. The greater annual production of monogamous females was offset by their shorter life expectancy.

Some minor hypotheses regarding the evolution of clutch size in birds include the suggestion that clutch size is the outcome of opposing predator and prey adaptations (Ricklefs, 1970) and the notion that selection for small, inconspicuous nests in the tropics accounts for the small clutch sizes that prevail there (Snow, 1978). The first is speculative, and the second fails to explain why those tropical species that build large, conspicuous nests also lay small clutches.

Age of First Reproduction and Semelparity

Students of the natural history of birds have suggested that the age of first reproduction was often postponed until the animal was mature enough either to establish a territory successfully or to find enough food for its young as well as for itself (Lack, 1954, 1966, 1968; Ashmole, 1963, 1971; Amadon, 1964; Cody, 1971). This hypothesis is perfectly compatible with the argument put forward earlier. Clutch size zero (nonbreeding) represents the smallest clutch size that a female can produce at a particular age and still replace herself and her mate, if reproducing even one young at that age is likely not only to be unsuccessful but also to reduce the female's probability of reproducing successfully in the future.

Although Cole (1954) showed that postponing reproduction results in reducing a population's rate of growth (r), Hamilton (1966), Mertz (1971a, b), and Giesel (1976) suggested that early reproduction is favored in increasing populations and late reproduction is favored in declining populations because delayed reproduction slows the rate of decline. This conclusion is certainly incorrect. The age of first reproduction is determined by the advantages accruing to the individual rather than to the population, and under any conditions those individuals capable of breeding earlier than others will be favored, simply because they will have the greatest ρ.

It may be so that the age of first reproduction is later in declining populations than in increasing populations, but this comes about by changing environmental conditions acting on individual success. When a population is small relative to its resources, young individuals may be successful in establishing territories or in finding sufficient quantities of food for their young, and thus the young may begin breeding successfully at an earlier age than under other conditions. As the population approaches its "steady-state" density, intraspecific competition prevents young individuals from establishing territories or reduces their ability to find sufficient food for their young, forcing the younger animals to postpone breeding until a later age. As the individual ρ values decline, so does the population's r. If the physical conditions deteriorate, the younger individuals and even older animals should postpone breeding until conditions improve if doing so increases their chances of increasing their lifetime reproduction compared with attempting to breed under the severe conditions. The maximum ρ and the population's r become negative, and the population declines. At every density and under all environmental conditions each individual is breeding at the earliest possible age consistent with maximizing its lifetime reproduction. Declining population size is a consequence, not a cause, of postponing reproduction.

With regard to whether a species is semelparous or iteroparous, Cole (1954) had shown that the advantage of immortality was offset by adding a single egg to the clutch of an individual that reproduces once and dies. Bryant (1971) reached the same conclusion with the more realistic assumption of including some adult and juvenile mortality. Subsequently, Charnov and Schaffer (1973) showed that the Cole and Bryant models were special cases. In the general case, the additional number of young that a semelparous species had to add to its clutch to gain the same rate of increase as an iteroparous species was not 1.0 but P/C individuals, where P is the annual adult survivorship and C is the proportion of young surviving the first year. When adult survivorship is high relative to juvenile survivorship, P/C is large, when it is low, P/C is small. Thus, it has been suggested, iteroparity will be selected when prereproductive survival is low compared with adult survival (Holgate, 1967; Murphy, 1968; Charnov and Schaffer, 1973; Schaffer, 1974a, b).

A correlation between low P/C ratios and the occurrence of semelparity has yet to be documented. Even if such a correlation should exist, it would not provide an explanation for the evolution of semelparity. As we have seen in the hypothetical, numerical example (Table 4.2), whether an individual is semelparous or iteroparous depends on the age of first reproduction relative to the age of last reproduction rather than on the shape of the survivorship curve. Pacific salmon, for example, are semelparous because the long migration upstream prevents successful reproduction by the younger, and smaller, age classes, not because of its adult/juvenile survivorship ratio.

CLUTCH SIZE AND POPULATION DYNAMICS

The conflicting views on the evolution of clutch size held by Lack (1954, 1966) and by Wynne-Edwards (1955, 1962) and Skutch (1967, 1976) affect these authors' hypotheses regarding population dynamics. Lack (1954, 1966) hypothesized that natural selection maximized the reproductive rate by adjusting the clutch to the size the parents could raise successfully. Parents that raised more were less successful because their young were underweight at fledging and had a low probability of survival. Parents that raised fewer were simply outbred. Because the reproductive rate is maximized, more young are produced than are necessary to maintain a "steady state." Lack (1954, 1966) hypothesized that natural populations are regulated by density-dependent mortality.

Wynne-Edwards (1955, 1962) and Skutch (1967, 1976) proposed instead that reproductive rates evolved in response to the populations' mortality rates. Populations of long-lived individuals evolve low reproductive rates, and populations of short-lived individuals evolve high reproductive rates. This adjustment of reproductive rates balances the mortality rates and prevents overpopulation. The theory explicitly invokes group selection, and therefore it has not been considered seriously by ecologists, who in general find group selection theories unacceptable (Lack, 1966; Wiens, 1966; Williams, 1966a; Wilson, 1975).

The theory of the evolution of clutch size proposed in this chapter offers a different interpretation. Selection results in the number of eggs per clutch and per year that allows the parents to replace themselves in the next generation. This, however, does not mean that the population is reproducing at the replacement rate. Returning to the hypothetical example (Table 4.2), the average m_x is 5.333 when breeding begins in the second year. Because eggs come in increments of 1, the minimum annual production of eggs is six. Thus, the actual clutch size exceeds the calculated replacement clutch size and results in a positive r, leading to population growth. Sooner or later, however, the population becomes limited in one of the ways described in Chapter 3. For example, as a food-limited population increases in size, intraspecific competition for food results in lowering the probabilities of survival (λ_x), and therefore raising the replacement clutch size just at the time intraspecific competition makes raising a larger brood more difficult. The population stops growing.

r-SELECTION AND K-SELECTION

In their treatise on island biogeography, MacArthur and Wilson (1967) contrasted the consequences of selection on uncrowded versus crowded populations. In the early stages of colonization, when populations are small and uncrowded, selection favors productivity. Those individuals that are able to harvest the most food (even wastefully) and raise the largest families will have an advantage in uncrowded habitats. As the habitat becomes crowded, selection favors efficiency. Those individuals that are able to harvest food efficiently in the face of competition when food is in short supply will have an advantage in crowded habitats. MacArthur and

Wilson (1967) termed the selection occurring in these extreme conditions "*r*-selection" and "*K*-selection," respectively. These ideas have become more general and now contrast the action of selection on populations occupying unstable, fluctuating, or ephemeral habitats with that on populations occupying relatively stable, persisting habitats. A high *r* value allows a species to discover a habitat quickly, to reproduce rapidly and use up the resources before other species can or before the habitat disappears, and to disperse in search of suitable habitats (Wilson, 1975). The consequences of "*r*-selection" and "*K*-selection" are many (Table 4.3). These include not only differences in population growth rate and colonizing ability but in survivorship, mode of mortality, length of life, social behavior, and body size.

In the ecological literature the symbol *r* has taken on several distinct meanings:

1. The symbol *r* represents the natural rate of increase of a population with constant age-specific death and birth rates and stable age distribution (Lotka, 1925). It is the difference between the birth rate per individual and the death rate per individual of the population but is usually calculated from Eq. (2.4). As such, *r* is a variable, having different values for conspecific populations exposed to different environmental conditions. When it is assumed for purposes of analysis that the prevailing environmental conditions are constant, l_x, m_x, and *r* will be constant, and the population's size will change exponentially [Eq. (2.1)]. Under such conditions, *r* has been called the "true," "standardized," or "stable rate of increase" (Dublin and Lotka, 1925), the "Malthusean parameter" (designated *m* by Fisher, 1930) for populations of genetically identical individuals, and the "ultimate rate of natural increase" (Mertz, 1970, 1971a).

The actual rate of increase of naturally occurring populations usually cannot be calculated in this way because they do not often reach a stable age distribution. Leslie (1945) provided a means for calculating a population's actual rate of increase regardless of its age distribution.

Andrewartha and Birch (1954) unfortunately called both this conception of *r* and the next the "intrinsic rate of increase" and symbolized them r_m. It seems better to distinguish between the "true rate of increase" *r* and the "intrinsic rate of increase" r_m, discussed next.

2. The symbol *r* is sometimes used to represent the maximum rate of increase that a population can achieve in an uncrowded environment under a specific set of physical conditions. This has been called the "incipient" (Lotka, 1927), "intrinsic" and "inherent" (Lotka, 1943), and the "innate" and "infinitesimal" rate of increase (Anderwartha and Birch, 1954). It is best symbolized r_m (Andrewartha and Birch, 1954).

When r is calculated for a population that is not only uncrowded but growing in an optimal physical environment, r_m becomes the maximum rate of increase, sometimes represented r_{max}, and it is the rate that is limited only by the inherent characteristics of the individuals of the population.

3. The symbol r also appears in the exponential and logistic equations of population growth, in which it is again a measure of the rate of increase of a population with constant age-specific birth and death rates and stable age distribution. In these equations, however, r is a constant, regardless of the population's density and, in the case of the logistic, the population's true rate of increase. Sometimes the r value used in these equations is r_m, but any value of r can be used.

Evans and Smith (1952) and Smith (1954) showed that high r_m values characterize species of small size and short generation times (Fig. 4.13). They suggested that r_m provides a measure of the average favorability or harshness of a population's environment and represents an evolutionary adjustment to the long-term conditions of the environment. An important result of Smith's (1954) analysis was his demonstration that the population's growth rate per generation R_0 is not a function of either the intrinsic rate of growth r_m or of the size of the individuals of the population. Therefore, populations of small animals do not necessarily grow any faster per generation than do populations of large animals, but populations of small animals with short generation times must have a high r_m to maintain the same growth rate per generation as populations of large animals with their longer generation times.

What, then, does it mean to say that a species is an "r strategist" or "K strategist"? Is r_m itself selected for? Or is r_m a consequence of selection operating on other aspects of the individuals of a population? Which of the correlates of "r-selection" and "K-selection" (Table 4.3) are cause and which are effect? I think that the concepts of "r-selection" and "K-selection" obscure rather than elucidate the cause and effect relations between the various population and life history parameters. The theory suggests that a population's rate of growth (r) evolves in response to the relative stability of the environment and intensity of competition. A rapid rate of population growth is specifically selected when a species occupies an unpredictable or ephemeral environment because a large r allows rapid occupation and exploitation of habitats as they become available. I believe this is a superficial explanation for several reasons.

1. Let us begin with the observations that generation time is related to body size (Fig. 4.11) and to the intrinsic rate of increase (Fig. 4.12): the smaller the body size, the shorter the generation time and greater the rate

of increase. I suggest that a small body size results in a short life expectancy and, therefore, short generation time, although why small body size necessarily means short life expectancy seems unknown. Furthermore, from the first law of population dynamics, we do know that the shorter the life expectancy, the greater the replacement rate (Table 2.6). Populations of individuals with short life expectancies must have greater birth rates than populations whose members have longer life expectancies, or they will not exist. The birth rate can be increased by breeding at an earlier age or by increasing fecundity. Whether earlier age of reproduction or greater fecundity is selected depends on the physiological characteristics of the species. A very small, short-lived species, such as a protozoan, whose potential fecundity is limited by its small size, has a high birth rate because of its short generation time. The minimal generation time of larger animals is limited by the length of time required for development to maturity, but fecundity can be great. If selection results in females producing only as many young as are necessary to replace themselves and their mates, we should expect short-lived species to be more fecund than longer-lived species, except as modified by the age of first reproduction, regardless of the relative stability of the environment.

Rain puddles are ephemeral habitats. It is not the low r_m of human populations that prevents human beings from colonizing rain puddles. Rather, it is their large size. Small size is a prerequisite for colonizing rain puddles; small size means short life expectancy; short life expectancy means high replacement rate; high replacement rate means a high birth rate (if the population exists); high birth rate means high r_m. Perhaps, instead of discussing "*r*-selection" and "*K*-selection," we should be discussing "birth selection" and "death selection" (Hairston *et al.*, 1970), or better, "small animals" and "large animals."

Further insight might be gained by imagining ourselves to be aphids, "small animals," who happen to be sufficiently intelligent to think about and discuss population dynamics. An aphidographer would soon develop the equations of A. J. Lotka, for they describe universal relationships. Other aphids would describe the statistics of l_x, m_x, r, R_0, and so on, of nonaphid populations and find, as did Smith (1954), that small animals have large r_m values and short generation times and large animals have low r_m values and long generation times. The aphidocentric aphids consider that their world is the best of all possible worlds. Environmental change is slight during the span of individual lifetimes, and even for several consecutive generations, compared with the magnitude of environmental changes to which larger animals are exposed. Aphids pass the periods unfavorable for feeding and reproducing as eggs. From the aphids' point of view, attempting to live actively during the entire year must be extraordinarily difficult for those animals that do, as evidence by their low fecundity. This low fecundity requires the evolution of long lifetimes; otherwise large animals would be extinct. We human beings probably would disagree with the aphids' interpretation, but I am sure the aphid ecologist would probably disagree with the usual human interpretation of its population dynamics and life history.

Furthermore, it is not insignificant that despite large differences in r_m the rates of increase per generation (R_0) of populations are of similar magnitude (Smith, 1954). To the aphid ecologist, their populations will not appear to be growing any more rapidly, or filling up their habitat any more rapidly, than our population appears to be growing to us (although this is alarming enough).

The concepts of "r-selection" and "K-selection" ignore both the time scale and the size scale in comparing the r values of different populations.

2. Another factor overlooked in evaluating the significance of r values of different populations is the length of the breeding season. In Chapter 2, the effect of intermittent breeding on the annual rate of population increase, r'_a, was discussed. Leslie and Ranson (1940) showed that for a given value of r, the annual rate of population increase was a function of the length of the breeding season. In their laboratory study of the vole (*Microtus agrestis*), r_m was 0.0877/week, generation time was 20.25 weeks, and R_0 was 5.904. If the vole bred throughout the year at this rate, the population would increase 95.6 times in 1 year. However, the vole is not a continuous breeder. How well the vole does from one year to the next depends on the length of the breeding season, as shown for the hypothetical example in Chapter 2 (Fig. 2.2 and Table 2.3). Populations with high r_m but short breeding seasons, then, may not do as well as populations with low r_m and long breeding seasons. Populations with R_0 much greater than 1.0, such as *Physa gyrina* with R_0 of 418 (DeWitt, 1954), *Tribolium castaneum* with 275 (Leslie and Park, 1949), *Calandra oryzae* with 113.56 (Birch, 1948), and others shown in Fig. 4.12, undoubtedly have either short breeding seasons or high mortality during the nonbreeding season.

It should be noted that r_m is a measure of the maximum rate of increase that a population can achieve in uncrowded conditions. It is usually measured in laboratory situations where optimal conditions are likely to be maintained sufficiently long to acquire the necessary data. Most animal populations reproduce seasonally, and what are lacking for understanding the significance of these r values are data on the length of the breeding season, the mortality rates during the nonbreeding season, and l_x and m_x schedules in natural conditions.

3. The so-called r-selected species are said to occupy "unstable" or

"unsaturated" habitats, whereas the "*K*-selected" species are supposed to occupy "stable" or "saturated" habitats. However, populations of "*r*-selected" species may well reach the "carrying capacity" of their habitats more rapidly and more often than a "*K*-selected" species because short-lived (e.g., seasonal or ephemeral) habitats provide the ultimate in crowding and extinction. The growth of such a population is J-shaped, a rapid increase followed by a rapid decline. The only reason that the population does not fluctuate indefinitely, or at least as long as those populations considered "persistent," is that the conditions favorable to survival and growth disappear.

The dynamics of populations that undergo large fluctuations is no different from that of populations undergoing small fluctuations. The difference is quantitative, not qualitative. The magnitude of the fluctuations of food-limited populations, for example, as we saw in Chapter 3, is a function of how rapidly the population consumes its food resource relative to the rate at which it is replenished.

The population dynamics models presented in Chapter 3 do not involve the logistic equation, which is considered inapplicable to naturally occurring or even laboratory populations, do not consider the *r* and *K* of the logistic as meaningful population parameters, and see no qualitative differences between (a) large and small populations, (b) rapidly and slowly growing populations, (c) founder and established populations, and (d) unstable and stable habitats. Thus, the concepts of *r*-selection and *K*-selection have no meaning within this framework of population dynamics. Furthermore, if natural selection favors females who are producing as few eggs or young as possible while still replacing themselves and their mates, as proposed earlier in this chapter, then different reproductive capacities between species or populations reflect probabilities of survivorship or the length of the breeding season of intermittently breeding populations (Chapter 2) or both. High fecundity or short generation time, which produce high r_m values, are adaptations to short life or brief breeding seasons rather than adaptations to colonize and exploit ephemeral, fluctuating, or unpredictable environments. Thus, for example, marine bottom invertebrates with long periods of planktotrophic pelagic larval life are characterized by high fecundity compared with those species with shorter periods of planktotrophic life, simply because the former have a lower probability of reaching reproductive age (Thorson, 1950). Again, the concepts of *r*-selection and *K*-selection have no meaning.

Clearly, however, some populations are capable of rapid growth. I consider this to be the result of selection for high fecundity because the individuals of these populations have a low probability of reaching reproductive age and have a short period favorable for breeding. If the periods of high fecundity and high mortality do not coincide, then during periods favorable for reproduction, rapid growth will occur (Fig. 4.17).

Population Consequences of Natural Selection

Natural selection refers to the changing frequencies of traits or alleles within a population, and population dynamics refers to the changing numbers of individuals of a population, both changes resulting from the effects of environment on the probabilities of surviving and reproducing. In a genetically diverse population, traits that maximize $\Sigma \lambda_x\mu_x$ and ρ will be selected and increase in frequency with respect to the alternative traits of other individuals. Selection for the maximum $\Sigma \lambda_x\mu_x$ and ρ occurs in all populations, whether they are increasing, decreasing, or maintaining their numbers.

Continual selection for individuals with the maximum ρ does not result in increasing the population's growth rate r simply because (1) the rate at which an allele or trait increases is not constant but varies with density as density affects individual probabilities to survive (λ_x) and reproduce (μ_x), and (2) under a particular set of conditions the maximum ρ is zero or even negative.

In Chapter 3 we saw how a population's survivorship declined with increasing density above the upper critical density or its age-specific birth rates declined after the available territories were filled. These changes come about because individual probabilities of surviving and reproducing decline as intraspecific competition for food, space, or other resource

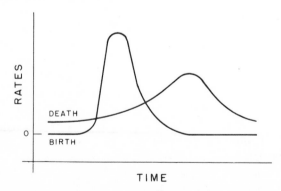

Figure 4.17. Seasonal variation in birth and death rates. When periods of high birth rates and high death rates do not coincide, a population's numbers will increase rapidly and then decline rapidly.

intensifies. Such is the case for any genotype. Nevertheless, selection for maximum $\Sigma \lambda_x \mu_x$ has consequences for population dynamics.

Consider first a food-limited population. Below the UCD, the population has a maximum growth rate determined by survivorship and fecundity, but, above the UCD, crowding leads to intraspecific competition for food, an increase in mortality, a decrease in fecundity, and, as a result, a decline in the growth rate (Fig. 3.5). Eventually, the population stops growing and fluctuates about its "steady-state" size. Any morphological, physiological, or behavioral change that increases individual efficiency in finding, capturing, transporting, consuming, or metabolizing food items should be selected for because these changes should reduce mortality and permit an increase in fecundity. As a consequence, the population's maximum growth rate below the UCD is greater, intraspecific competition for food begins at a higher density (i.e., the UCD is greater), and the population can reach a larger "steady-state" size (Fig. 4.18).

Food-limited populations in nature should be rare, however, because any persistent source of mortality, say from predation, parasites, or disease, is added to crowding effects, lowering the population's survivorship and,

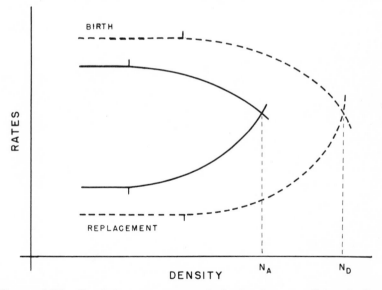

Figure 4.18. Population consequences of selection for more efficient feeding. The birth and replacement rates of the ancestral food-limited population are drawn with solid lines. The same rates for the descendant population are drawn with dashed lines. With more efficient abilities in finding, capturing, transporting, or metabolizing food, the replacement rate decreases, the birth rate increases, the UCD occurs at a greater density, and the population is able to achieve a larger "steady-state" population.

perhaps, fecundity, at every density. The population's maximum growth rate is lower at every density, and the "steady-state" size is smaller than it would be without mortality from predation, parasitism, or disease. This point is taken up in greater detail in the next chapter, where a numerical example is provided (Table 5.2; Fig. 5.3). Selection for greater crypticity, for example, should raise the probabilities of individual survival, increasing the population's $\Sigma \, l_x$ and reducing its replacement rate. The reduction in mortality from predation results in a greater growth rate at all densities and in a larger "steady-state" size (Fig. 4.19).

In space-limited populations, selection for increased survival and fecundity will not affect the size of the breeding population, if the size of the minimal tolerable territory remains unchanged. Natural selection can affect survivorship, replacement rate, and the stable age structure. Assuming that the "floaters" do not disperse or that emigration is balanced by immigration, the population's total size can increase, but the frequencies of the groups (young, breeding adults, and "floaters") will change, even though the number of territory holders remains unchanged. If the additional "floaters" do disperse, they are likely to be recorded as deaths, and the additional adult survivorship or increased fecundity will appear to be

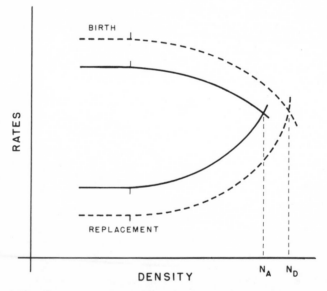

Figure 4.19. Population consequences of selection for cryptic coloration and behavior. The birth and replacement rates of the ancestral population are drawn with solid lines. The same rates for the descendant population are drawn with dashed lines. As cryptic coloration and behavior reduce predation, the population's replacement rate decreases, birth rate increases, and maximal "steady-state" population size increases.

"balanced" by increased mortality rates among the young. For the breeding population to increase in size, selection must result in a smaller tolerable territory size. Selection, however, seems to favor individuals who can establish the largest territories allowed by their time and energy budgets (Brown, 1964). At least early birds often establish larger territories than they finally maintain when later birds arrive (Hinde, 1956; Brown, 1969a), and some birds establish larger territories than they subsequently use in acquiring food (Lanyon, 1956; Stefanski, 1967; Zimmerman, 1971).

Postponing the age of first reproduction, while increasing the individual's $\Sigma \lambda_x \mu_x$ and ρ relative to those of alternative life history patterns under particular conditions, always results in reducing the population's potential rate of growth (r). If there are no other changes in the survivorship and fecundity schedules, the population's maximum size is also reduced (Fig. 4.20). The evolution of delayed breeding places a population's survival in jeopardy should the population be exposed to a climatic trend toward more severe conditions or to a greater intensity of predation, such as is imposed by the human species on other long-lived but low-fecundity species, such as whales.

These examples of the population consequences of selection for different survivorship and fecundity schedules assume only a single change, but life is no doubt more complex. We can imagine a more complex series of interactions. Should larger body size evolve as a means of reducing the

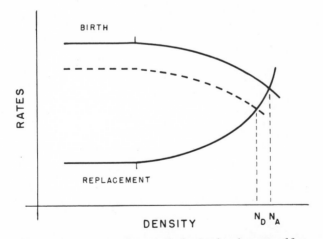

Figure 4.20. Population consequences of selection for a later age of first reproduction. The birth and replacement rates for the ancestral population are drawn with solid lines. The birth rate for the descendant population is drawn with a dashed line. Mortality is unaffected, and therefore the replacement rate is unchanged. Postponing the age of first reproduction lowers the population's birth rate and its maximal "steady-state" size.

effects of predation, then the population's growth rate and "steady-state" size should increase, as indicated above (Fig. 4.19). However, larger body size increases individual demand on the available food supply, and intraspecific competition begins at a lower population density, reducing the population's potential maximum size. Larger body size, however, may lead to longer life, increasing the population's maximum growth rate and population size, but larger size may result in a longer development time to reach maturity, postponing breeding and lowering the population's maximum growth rate and population size. Finally, with larger body size, the population's available food supply changes in ways that cannot be predicted. Former prey may no longer be suitable, but formerly inappropriate prey may now be taken. The relative abundance of prey species cannot be known beforehand, but it will affect the size of the population.

The consequences of individual selection for population dynamics in nature are no doubt multiple and, therefore, complex. Nevertheless, models of the effects of single changes are useful, and it is interesting that a single change in the survivorship or fecundity schedule in food-limited and predator-limited populations affects both the population's naximum potential growth rate (r) and its maximum potential "steady-state" size.

5

PREDATION AND MAXIMUM SUSTAINABLE YIELD

The concept of maximum sustainable yield has guided recommendations for the management of resources for several decades. So few resources have been managed successfully to sustain a maximum yield that the concept has fallen into disrepute (Eberhardt, 1977; Larkin, 1977; Talbot, 1977). Indeed, Larkin (1977) has written its epitaph:

M. S. Y.
1930s–1970s

Here lies the concept, MSY.
It advocated yields too high,
And didn't spell out how to slice the pie.
We bury it with the best of wishes,
Especially on behalf of fishes.
We don't know yet what will take its place,
But hope it's as good for the human race.

R. I. P.

With the increasing pressure of human populations on the world's resources, societies will either rationally manage their resources or suffer the consequences of overharvesting. Whether the management concept used in recommending policy is called maximum sustainable yield or optimal sustainable population (Eberhardt, 1977; Larkin, 1977) or some other epithet, the dynamics of predation on resource populations remains unchanged. Any management effort requires an understanding of the dynamics of predation.

Predator–prey relationships have received considerable attention in the ecological literature, but only Slobodkin (1961, 1968, 1974) seems to have addressed the problem of how a predator should go about consuming its prey without damaging the prey population's ability to replace itself and continue to serve as a food supply for the predator. According to Slobodkin, the so-called prudent predator should take only those individuals that have a high probability of dying from other causes. Seemingly, however, prudent predation can occur only at low yields (Slobodkin, 1968; Mertz and Wade, 1976). Furthermore, Maynard Smith and Slatkin (1973) and Maiorana (1976) pointed out that a prudent predator that refrained from taking available prey could evolve only by group selection, but Slobodkin (1974) suggested that the apparent prudence of natural predators was a result of the prey's adaptations, which make it difficult for predators to be imprudent.

In fisheries management, the analysis of dynamics takes a different form and reaches different conclusions. "Logistic-type models" (Graham, 1935; Schaefer, 1968) recommend that in order to maximize a sustainable yield a predator should maintain the prey below the unharvested steady-state population because the population's rate of growth is greatest at one-half the maximum density. It has been suggested, however, that high fecundity animals can sustain a maximum yield when fished well below the 50% level, but low fecundity animals may not be able to sustain a yield when fishing reduces the population much below its unharvested steady-state size (Eberhardt and Siniff, 1977).

"Dynamic pool models" (Beverton and Holt, 1957; Schaefer, 1968) relate the yield per recruit in steady-state populations being subjected to different rates of fishing mortality. They do not determine whether a particular fishing rate will result in prey population decline, but instead assume that compensatory changes in mortality and reproduction during the prerecruit phase maintain a constant recruitment rate into the fishery (Beverton and Holt, 1957; Gulland, 1962; Schaefer, 1968).

None of these ecological and management models analyzes population dynamics in terms of the parameters of Lotka's equations—l_x, m_x, r, b, d, and c_x. Any change in predation affects the values of each of these parameters, and whether a population can sustain any yield, much less a maximum yield, depends on how the statistical values are changed with different patterns of predation. We need a model that relates different patterns of predation to Lotka's parameters.

EFFECTS OF PREDATION ON PREY POPULATIONS

Recall the dynamics of a food-limited population unexposed to predation (Fig. 3.5). The population grows at a maximum rate until the UCD is reached. Then the replacement rate increases, the birth rate decreases, and population growth slows until it stops. Above the UCD, the replacement rate increases because increasing intraspecific competition for food results in greater mortality or dispersal. The shape of the survivorship curve depends on the age-specific mortality rates (Fig. 2.2). Changes in the birth rate above the UCD may be caused by the change in age structure resulting from mortality, dispersal, or both, or by a decline in individual fecundity, or some combination. Two survivorship schedules characterize this population: maximal survivorship during the growth phase below the UCD and minimal survivorship at "steady state." Intermediate survivorship schedules occur during the transition from the growth phase to the "steady-state" phase, but these need not be considered.

Any predation decreases the probability of surviving the ages exposed to predation, increasing age-specific death rates (q_x) and decreasing survivorship (Table 5.1 and Fig. 5.1), which in turn result in increasing the replacement rate (Table 5.1). Furthermore, predation lowers the frequencies of the preyed-upon age classes (Table 5.1), and the changed age structure lowers the prey population's birth rate, provided that m_x values remain unchanged (Table 5.2).

Predation, then, affects a population's death rate, birth rate, and age distribution. Whether a population can sustain a "steady state" in response to predation depends on whether its actual birth rate exceeds or equals its replacement rate. Should the birth rate fall below the replacement rate or the replacement rate rise above the birth rate, the prey population is being overharvested, and it will decline to extinction.

The permutations of age-specific birth rates, age-specific predation rates, and replacement rates are so numerous that no simple equation can give us the predation rate that will allow a sustainable maximum yield of a food-limited prey population. Several hypothetical numerical examples can illustrate the dynamics of predator–prey relationships and perhaps allow us to pose some generalizations.

The minimum data required are the maximum survivorship schedule (below the UCD), the minimal survivorship schedule with crowding at "steady state," and the age-specific birth rates (m_x) in the absence of crowding (below the UCD) of the unharvested population. To these data is added mortality from proposed or actual predation. The new birth and replacement rates are calculated and compared.

TABLE 5.1

Survivorship, Mortality, and Age Structure of Several Populations Exposed to Different Patterns of Predation[a]

	Age class										
Population	1	2	3	4	5	6	7	8	9	10	Replacement rate
A											
l_x	1.000	0.900	0.800	0.700	0.600	0.500	0.400	0.300	0.200	0.100	0.182
q_x	0.100	0.111	0.125	0.143	0.167	0.200	0.250	0.333	0.500	1.000	
c_x	0.182	0.164	0.146	0.127	0.109	0.091	0.073	0.055	0.036	0.018	
B											
l_x	1.000	0.800	0.711	0.622	0.533	0.444	0.355	0.267	0.178	0.089	0.200
q_x	0.200[b]	0.111	0.125	0.143	0.167	0.200	0.250	0.333	0.500	1.000	
c_x	0.200	0.160	0.142	0.124	0.107	0.089	0.071	0.053	0.036	0.018	
C											
l_x	1.000	0.700	0.622	0.545	0.467	0.389	0.311	0.233	0.156	0.078	0.222
q_x	0.300[b]	0.111	0.125	0.143	0.167	0.200	0.250	0.333	0.500	1.000	
c_x	0.222	0.155	0.138	0.121	0.104	0.086	0.069	0.052	0.035	0.017	
D											
l_x	1.000	0.600	0.533	0.467	0.400	0.333	0.267	0.200	0.133	0.067	0.250
q_x	0.400[b]	0.111	0.125	0.143	0.167	0.200	0.250	0.333	0.500	1.000	
c_x	0.250	0.150	0.133	0.117	0.100	0.083	0.067	0.050	0.033	0.017	
E											
l_x	1.000	0.900	0.800	0.700	0.600	0.500	0.350	0.228	0.129	0.052	0.190
q_x	0.100	0.111	0.125	0.143	0.167	0.300[b]	0.350[b]	0.433[b]	0.600[b]	1.000[b]	
c_x	0.190	0.171	0.152	0.133	0.114	0.095	0.067	0.043	0.025	0.010	
F											
l_x	1.000	0.900	0.800	0.700	0.460	0.291	0.175	0.096	0.045	0.014	0.223
q_x	0.100	0.111	0.125	0.343[b]	0.367[b]	0.400[b]	0.450[b]	0.533[b]	0.700[b]	1.000[b]	
c_x	0.223	0.201	0.178	0.156	0.103	0.065	0.039	0.021	0.010	0.003	
G											
l_x	1.000	0.900	0.800	0.700	0.390	0.208	0.104	0.047	0.017	0.003	0.240
q_x	0.100	0.111	0.125	0.443[b]	0.467[b]	0.500[b]	0.550[b]	0.633[b]	0.800[b]	1.000[b]	
c_x	0.240	0.216	0.192	0.168	0.094	0.050	0.025	0.011	0.004	0.001	

[a] Survivorship curves shown in Fig. 5.1.
[b] Preyed-upon age classes.

Consider a population whose survivorship when uncrowded and unharvested is given by schedule UU in Table 5.2 and whose crowded but unharvested "steady-state" survivorship is given by schedule UC. In this population, crowding increases the mortality of only the youngest age class (from 5% when uncrowded to 15% when crowded), and crowding reduces m_x from 0.5 for all reproductive classes (ages 4–10) to 0.231 at "steady state," assuming, for simplicity, that crowding affects the fecundity of all age classes equally.

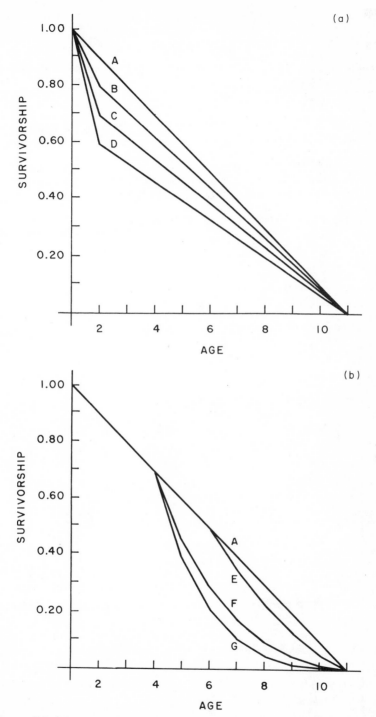

Figure 5.1. Survivorship curves under different predation strategies. The schedules are given in Table 5.1.

TABLE 5.2

Effects of Different Predation Intensities on a Prey Species' Population Parameters

	Age class										b	b_r	r
	1	2	3	4	5^a	6^a	7^a	8^a	9^a	10^a			
Unharvested, uncrowded													
UU q_x	0.050	0.053	0.056	0.059	0.063	0.067	0.071	0.077	0.167	1.000	0.239	0.130	0.138
UU l_x	1.000	0.950	0.900	0.850	0.800	0.750	0.700	0.650	0.600	0.500			
UU m_x	0.000	0.000	0.000	0.500	0.500	0.500	0.500	0.500	0.500	0.500			
Unharvested, crowded, "steady state"													
UC q_x	0.150	0.053	0.056	0.059	0.063	0.067	0.071	0.077	0.167	1.000	0.143	0.143	0.000
UC l_x	1.000	0.850	0.805	0.760	0.715	0.670	0.625	0.581	0.536	0.446			
UC m_x	0.000	0.000	0.000	0.231	0.231	0.231	0.231	0.231	0.231	0.231			
Survivorship of uncrowded (U) and crowded (C) populations harvested at the indicated frequency													
10U l_x	1.000	0.950	0.900	0.850	0.800	0.670	0.558	0.463	0.381	0.279	0.232	0.146	0.112
10C l_x	1.000	0.850	0.805	0.760	0.715	0.598	0.498	0.413	0.340	0.249	0.161	0.161	0.000
20U l_x	1.000	0.950	0.900	0.850	0.800	0.590	0.433	0.315	0.228	0.114	0.227	0.161	0.089
20C l_x	1.000	0.850	0.805	0.760	0.715	0.527	0.386	0.281	0.203	0.128	0.177	0.177	0.000
30U l_x	1.000	0.950	0.900	0.850	0.800	0.510	0.323	0.203	0.126	0.067	0.221	0.175	0.065
30C l_x	1.000	0.850	0.805	0.760	0.715	0.455	0.288	0.181	0.113	0.060	0.191	0.191	0.000
40U l_x	1.000	0.950	0.900	0.850	0.800	0.430	0.229	0.121	0.063	0.027	0.215	0.186	0.044
40C l_x	1.000	0.850	0.805	0.760	0.715	0.331	0.176	0.093	0.049	0.021	0.208	0.208	0.000
50U l_x	1.000	0.950	0.900	0.850	0.800	0.350	0.152	0.065	0.027	0.009	0.212	0.196	0.024
50C l_x	1.000	0.850	0.805	0.760	0.715	0.312	0.135	0.058	0.025	0.008	0.214	0.214	0.000
60U l_x	1.000	0.950	0.900	0.850	0.800	0.270	0.090	0.030	0.010	0.002	0.207	0.204	0.005
60C l_x	1.000	0.850	0.805	0.760	0.715	0.241	0.080	0.026	0.008	0.002	0.223	0.223	0.000
63U l_x	1.000	0.950	0.900	0.850	0.800	0.246	0.074	0.022	0.007	0.001	0.206	0.206	0.000

[a] Preyed-upon age classes. The rate is indicated in left-hand column.

Let us now harvest age classes 5–10 at some constant rate. The age-specific mortality of the preyed-upon age classes increases, and this is reflected by decreasing survivorship through these age classes (Table 5.2). With increasing intensity of predation, the population's birth rate decreases and replacement rate increases (Table 5.2, Figs. 5.2 and 5.3). By plotting these values, the "steady-state" birth and replacement rates and population size can be determined from the point of intersection of the birth and replacement rate curves (Fig. 5.3). This population can sustain a

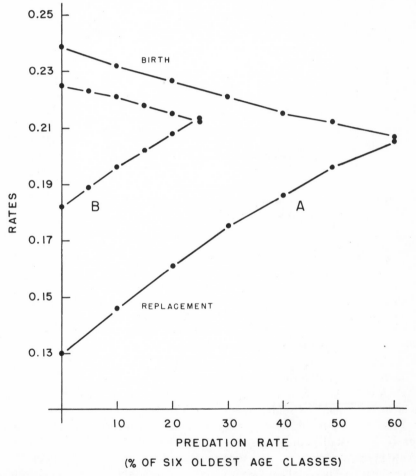

Figure 5.2. Birth and replacement rates of two uncrowded populations at different rates of predation. (A) Survivorship and reproductive schedules of population A are given in Table 5.2. (B) The survivorship schedule of population B is given as population A in Table 5.1. The reproductive schedule of population B is the same as that given in Table 5.2.

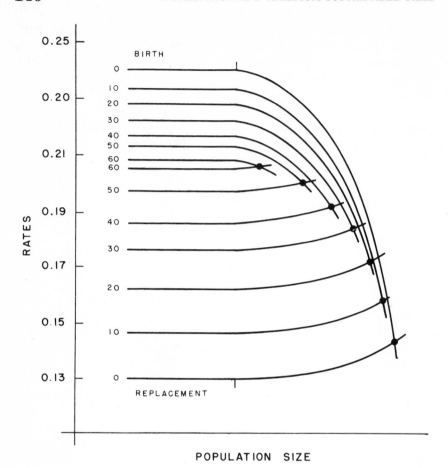

Figure 5.3. Birth and replacement rates for a food-limited population (predation rate 0) that is harvested at 10, 20, 30, 40, 50, and 60% of the six oldest age classes (see Table 5.2). The "steady-state" population size is determined by the intersection of the birth and replacement curves for a particular predation rate.

"steady state" when the predation rate is 63% on age classes 5–10 (Table 5.2). Any further predation will result in extinction because the birth rate will be less than the replacement rate.

From this example, we may suggest some conclusions. (1) The "steady-state" harvested populations are smaller than the "steady-state" unharvested population; the greater the predation rate the smaller the population (Fig. 5.3). (2) The smallest "steady-state" population, that is, the smallest population that can sustain itself despite predation, is at the UCD density of the unharvested population. (3) As predation intensifies, the prey population's maximum growth rate (r) declines toward zero, even as the prey population's size approaches uncrowded conditions.

This example illustrates the dynamics of predation and gives us a means of determining whether a particular predation strategy can be sustained by the prey population.

Effects of UCD, Survivorship, and Fecundity

In the growth model for food-limited populations presented in Chapter 3, two populations with identical l_x and m_x schedules can reach different sizes because their UCD points occur at different densities (Fig. 3.5). Nevertheless, because the l_x and m_x schedules are the same, the effects of a particular predation strategy on the values of birth and replacement rates are the same. The dynamics of predation on prey populations is independent of the prey population's size. When the maximum sustainable predation rate is imposed, both large and small populations are reduced to their UCD size.

Populations with lower survivorship (e.g., schedule A in Table 5.1) but the same m_x as the population just analyzed cannot sustain such high rates of predation (Fig. 5.2B), and populations with greater fecundity but the same survivorship can sustain higher rates of predation. Thus, survivorship and fecundity can affect the maximum rate of predation a population can withstand; the greater the survivorship and the greater the fecundity, the greater the sustainable predation rate. There is, however, an inverse relationship between survivorship and fecundity. Whether a high-survivorship, low-fecundity population can sustain a greater predation rate than a low-survivorship, high-fecundity population can be determined only by calculation. Furthermore, the "steady-state" size of harvested populations is independent of survivorship or fecundity because the UCD is independent of either survivorship or fecundity. A low-fecundity, high-survivorship population with a high UCD will have a larger harvested "steady-state" size than a high-fecundity, low-survivorship population with a lower UCD.

Yield

The yield in numbers per individual of the population is given by

$$Y_a = \Sigma \, h_x c_x, \tag{5.1}$$

where h_x is the age-specific harvest rate and c_x is the proportion that age class x represents in the harvested population (Table 5.3). This is a

TABLE 5.3

Per Capita Yield of Harvested Population

x	l_x	c_x	h_x	$c_x h_x$
1	1.000	0.20600		
2	0.950	0.19570		
3	0.900	0.18540		
4	0.850	0.17510		
5	0.800	0.16480	0.63	0.10382
6	0.246	0.05068	0.63	0.03193
7	0.074	0.01524	0.63	0.00960
8	0.022	0.00453	0.63	0.00285
9	0.007	0.00144	0.63	0.00091
10	0.001	0.00021	0.63	0.00013
				0.14924

sustainable harvest when $r \geq 0$ and when no age class is eliminated. In the above example, a yield could be sustained when a predator took 63% of the oldest six age classes (Table 5.2 and Fig. 5.3). Per capita yield of this population is 0.14924.

Although this yield is a sustainable yield, it is not necessarily a maximum sustainable yield. Is it possible to increase the yield by using a different predation strategy? Which predation strategy produces the maximum per capita sustainable yield can be determined only by trial and error. Additional strategies were considered 55% predation on the oldest seven age classes and 70 and 80% predation on the oldest five age classes. The first alternative strategy increases per capita yield to 0.180 but none of these strategies can be sustained (Table 5.4). The first because $r < 0$, and the others because age class 10 is virtually eliminated, few individuals reaching that age because of the heavy predation on the earlier age classes. For our example, 63% predation on the six oldest age classes appears to provide a maximum per capita sustainable yield, but other combinations must be tried before we can be certain. We can continue to use the numerical example (Table 5.2 and Fig. 5.3) for illustrative purposes by assuming $N_K = 8000$, UCD $= 5000$, and the degree of depression of survivorship is directly correlated with population size above the UCD. The birth rate, b, and the harvested "steady-state" size, N_H, can be taken directly from the graph (Fig. 5.3). New survivorship and age structure schedules must be calculated, and from these each population's per capita yield can be determined (Fig. 5.4).

The total sustainable yield (Y_t) of the prey population is the product of

TABLE 5.4

Life Tables of Two Overharvested Populations

x	Natural q_x	h_x	l_x	m_x	$l_x m_x$	$x l_x m_x$	h_x	l_x	m_x	$l_x m_x$	$x l_x m_x$
1	0.050		1.000					1.000			
2	0.053		0.950					0.950			
3	0.056		0.900					0.900			
4	0.059	0.55	0.850	0.5	0.425	1.700		0.850	0.5	0.425	1.700
5	0.063	0.55	0.332	0.5	0.166	0.830		0.800	0.5	0.400	2.000
6	0.067	0.55	0.129	0.5	0.065	0.387	0.8	0.750	0.5	0.375	2.250
7	0.071	0.55	0.049	0.5	0.025	0.172	0.8	0.100	0.5	0.050	0.350
8	0.077	0.55	0.019	0.5	0.010	0.076	0.8	0.013	0.5	0.007	0.056
9	0.167	0.55	0.007	0.5	0.004	0.032	0.8	0.002	0.5	0.001	0.009
10	1.000	0.55	0.002	0.5	0.001	0.010	0.8	0.000	0.5	0.000	0.000
			4.238		0.696	3.207		5.365		1.258	6.365
			$b_r = 0.236$					$b_r = 0.186$			
			$R_0 = 0.696$					$R_0 = 1.258$			

the prey's harvested "steady-state" size and the per capita sustainable yield:

$$Y_t = N_H \, \Sigma \, c_x h_x, \qquad (5.2)$$

where N_H is the harvested "steady-state" size, which is the population supporting the predation. This is a sustainable harvest so long as $r \geqslant 0$ and predation does not eliminate an age class.

The residual population size (N_R) is the harvested, "steady-state" size less the yield,

$$N_R = N_H - N_H \, \Sigma \, c_x h_x. \qquad (5.3)$$

Yields for the various predation rates in our illustrative example (Table 5.2 and Fig. 5.3) are shown in Fig. 5.4. In this population the per capita yield increases with increasing predation rate and is greatest at the maximum sustainable predation rate, 63%; but the maximum sustainable yield occurs at a 50% predation rate.

Maximal Sustainable Yield in Biomass

At N_K, the prey population is maximizing the conversion of its food resource into prey biomass and thus provides the greatest food resource for its predators. *No predation strategy can increase the prey population's biomass.* Any predator that reduces the biomass of its food resource

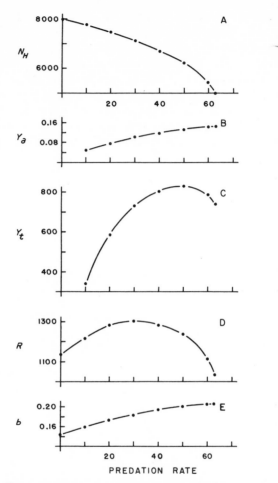

Figure 5.4. Changes in population parameters with increasing predation (survivorship and fecundity schedules in Table 5.2). (A) Harvested "steady-state" size, N_H. (B) Per capita yield, Y_a. (C) Total sustainable yield in numbers, Y_t. (D) Recruitment of age class 1 in numbers, R. (E) Birth rate, b.

necessarily curtails its own maximum population size. We must ask, how-ever, whether the strategy that maximizes yield in numbers also maximizes the yield in biomass, especially for those species in which density-dependent differences in individual growth rates exist.

In species of birds and mammals in which adult size is reached rapidly, changes in body size of the age classes exposed to predation are negligible.

Thus, maximizing the yield in numbers also maximizes the yield in biomass. In some other species, such as fishes, individual size often increases throughout life, and the rate of growth may be a function of population density (Fig. 5.5). The relationship between yield in numbers and yield in biomass is not immediately clear. At the UCD, individual growth rates will be at a maximum, but the MSY in numbers may occur at a lower predation rate on a larger population (Fig. 5.4). The MSY in numbers, then, need not be the MSY in biomass. Which predation rate produces the MSY in biomass can be determined only by trial and error.

Per capita yield in biomass (Y_b) is given by

$$Y_b = \Sigma \, c_x h_x w_x, \tag{5.4}$$

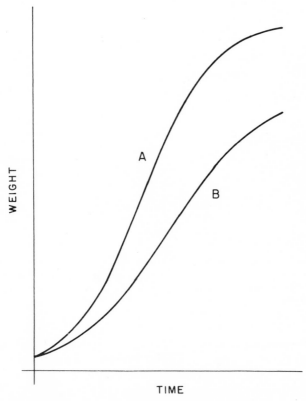

Figure 5.5. Growth of individual animals in populations of different density. (A) Below UCD. (B) At "steady state."

where w_x is the average weight of individuals of age class x. The total yield in biomass (Y_B) is given by

$$Y_B = N_H \Sigma\, c_x h_x w_x, \tag{5.5}$$

In the example, the maximum predation rate that produces a sustained yield is 63%, but the MSY in numbers is provided by 50% predation (Fig. 5.4). Inasmuch as individual growth rates should be greater at lower densities, the average weight of prey should increase as populations become smaller with increased predation. Thus, the MSY in biomass must occur between 50% and 63% predation rate. Which predation rate provides the MSY in biomass requires knowing the growth rates of individuals at N_H.

There is a danger of overharvesting when preying on species whose rate of growth of body size varies with density. For example, we plan to take 63% of age classes 5–10 of the crowded, unharvested population UC in Table 5.2. Our demographic data indicate that such a predation rate can be sustained by the prey population. However, by harvesting this population, we reduce its population size to its UCD level. As a result, individual growth rates increase, and our fishing gear now takes fishes of younger age classes, depressing the birth rate and increasing the replacement rate to greater extents than our calculations allowed. Thus, the gear, designed to take fishes of particular ages under one set of assumptions must now be redesigned to allow the younger but now bigger fishes to get away.

Recruitment

There are two populations involved in these calculations: the population without predation (N_K) and the population with predation (N_H). The latter is the observable population, and the characteristics of this population vary with the intensity of predation. As already noted, the harvested population decreases and the per capita yield increases with increasing predation (Fig. 5.4). The birth rate, interpolated from Fig. 5.3, increases with increasing predation, reaches a maximum at the maximum sustainable predation rate, and falls off with greater predation (Fig. 5.4E). Recruitment of age class 1 into the population is the product of the birth rate and the harvested population size. Recruitment in these populations increases with increasing predation but soon declines, the maximum recruitment occurring at 30% predation (Fig. 5.4D). Both the changes in the birth rate and recruitment are consequences of differences in each population's age structure resulting from the pattern of predation.

COMPLICATING FACTORS

"Compensation"

This analysis shows that, as the intensity of predation on an unharvested, "steady-state," food-limited prey population increases, not only does the prey population decrease in size, but its growth rate at any density is reduced. At *all* densities, even below the UCD, predation reduces the birth rate and increases the replacement rate in comparison with those rates in the unharvested population. Above the UCD, crowding effects further depress the birth rate and raise the replacement rate. The increases in the birth rate and survivorship observed in harvested populations result from the reduction in crowding effects as the prey population decreases. Under no condition of predation does the birth rate reach the birth rate of the unharvested population at the same density (Fig. 5.3).

Although up to this point we have considered only a food-limited population, the model is easily applicable to other situations. For example, as density increases, adults may begin cannibalizing the young, lowering survivorship of the young and increasing the population's replacement rate. Thus, the UCD is the point at which eating conspecific young becomes nutritionally beneficial for the adults. Again, if the predator removes adults, the population's birth rate and survivorship increase, but in this case it results from the reduction in cannibalism. Nevertheless, at a given density the birth rate is less than, and the replacement rate is greater than, the same rate without predation.

The notion of "compensation," that prey populations can or do respond to predation by increasing fecundity or by increasing the survivorship of survivors (Ricker, 1954; Beverton and Holt, 1957; Gulland, 1962; Schaefer, 1968), is not compensation at all but rather the result of a smaller population size and a reduction in crowding effects, whatever they may be. In the population growth models presented here (Fig. 3.5), l_x and m_x are maximal below the UCD. Predation cannot increase these values.

Other Predators

Often two or more predators feed on a single prey population, or more likely two or more predators feed on several species in a complex web of trophic relationships. Complex relationships are no different from simple relationships except in the even greater difficulty in determining predation rates on natural populations. In theory it is easy enough to determine the consequences of harvesting by two or more predators. The age-specific harvesting rate (h_x) is the combined effect of all predators feeding on a particular age class $(h_x{}^a + h_x{}^b + \cdots)$. Whether the combined predation by two or more predators can be sustained by the prey population is determined in the manner described above.

In complex relationships, one of the predators can increase its portion of the yield by eliminating the other predator(s), either by preying directly on the other predator(s) or through competition. For example, imagine predator species A and species B consuming common prey species. If species A also preys on species B, harvesting rates of species B on the prey population $(h_x{}^b)$ are reduced, allowing either the prey populations to increase or predator A to increase its harvesting rates $(h_x{}^a)$ on the prey species. Determining the sustainable yield of a multispecific fish community for the human predator is a challenging problem, requiring much detailed information on the dynamic and trophic relationships of all interacting members of the community.

Human beings are also competitors with other fish predators. This is illustrated in the recently (since 1956) developing anchovy fishery off the coast of Peru, where large populations of fish-eating birds breed on islands noted for their accumulations of guano. The bird populations are at the top of a food chain, which begins in the cold, upwelling waters of the Humboldt Current. Approximately every seven years a warm, southward-flowing current, known as El Niño, extends into the region, killing off the plankton, schooling fish, and birds (Murphy, 1936). A population of some 30 million birds declined to about 4 million in 1957, just as the anchovy fishery began to be developed (Fig. 5.6; Schaefer, 1970). Both the bird populations and the fishery increased during succeeding years, but the birds tapered off at about 18 million and crashed to 3 million during the El Niño of 1965, a population low from which they have not recovered (Paulik, 1971; Idyll, 1973). The anchovy fishery continued to increase and exceeded the recommended maximum sustainable yield. The combined yield to both avian and human predators had exceeded the recommended maximum sustainable yield earlier. It seems likely, although the evidence is circumstantial, that the bird populations are unable to obtain sufficient food for growth to their former numbers because of competition with the human predator.

Social Species

Many prey are social species in which males establish dominance hierarchies, which determine which males have access to space, food, and females. Such animals are crocodilians, seals, many ungulates, and gamebirds. The strategy of preying on the oldest and largest animals in order to maximize a sustainable yield may be disastrous when applied to social species because the elimination of the oldest, largest, and probably

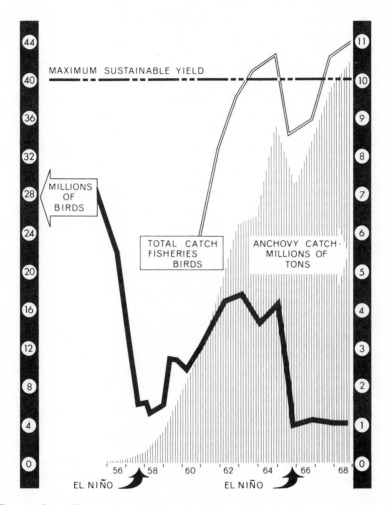

Figure 5.6. The numbers of fish-eating birds and the development of the anchovy fishery off the coast of Peru. As the anchovy fishery increased its activity, the bird populations declined. The bird population crashes are owing to the intrusion of the warm water current, El Niño, but the populations have not returned to their usual highs between the crashes, presumably because of competition with human predators. (From Jehl, 1970.)

dominant individuals leads to disruption of the dominant–subordinate relationships. The ensuing contest among the survivors for dominance can lead to reduced reproductive success because of the males paying insufficient attention to the females and the inadequate protection of the young.

Variations of l_x and m_x

Theoretical models have the benefit of assuming that the numerical values of parameters are constant. In nature, numerical values vary because of changing environmental conditions. Age-specific mortality and reproduction are likely to vary in time. This results in changes in annual recruitment (of age class 1 or of animals entering the fishery) and sometimes produces a year class that predominates the age distribution for many years (Lawler, 1965; Odum, 1971). As a consequence, a populations' size will change with time because its age distribution never stabilizes. In such a fluctuating population, the total yield will vary with population size. Three strategies for harvesting fluctuating populations are (1) to maintain a constant yield by increasing effort when populations are small and decreasing effort when populations are large, (2) to maintain constant harvesting rates by reducing effort when populations are small and increasing it when large, and (3) to determine average l_x and m_x values of the population, to determine the amount of effort to harvest a population with these values, assuming them to be constant, and to maintain this effort regardless of population size, overharvesting when survivorship and reproduction are low and underharvesting when survivorship and reproduction are high.

Ricker (1958), in analyzing numerical models based on his stock-recruitment curves of simple single-age prey species, concluded that the most effective management of a fishery for long-term maximum yield in fluctuating environments results in large variations in the catch and might even require completely closing a fishery periodically. Furthermore, the best constant rate of exploitation of fluctuating prey populations is the same as or close to the best rate of exploitation under constant conditions. Maintaining a constant rate requires varying fishing effort and results in varying yields.

In an analysis of a simple system assuming logistic growth, Beddington and May (1977) studied the consequences of environmentally imposed disturbances on a population being harvested to provide a sustainable yield. The time required to recover equilibrium conditions after disturbance is greater as harvesting effort and yield increase, and the predictability of the yield decreases as effort increases. These effects are greater when yield rather than effort is being held constant, suggesting that maintaining a constant effort is the preferred strategy of exploitation when environmental conditions are fluctuating.

Which strategy is best to exploit fluctuating populations seems to require further investigation.

LABORATORY EXPERIMENTS

Predation rates have been imposed on laboratory populations of the Australian sheep blowfly (*Lucilia cuprina*) (Nicholson, 1954a,b, 1958a), *Tribolium confusum* (Watt, 1955), *Daphnia pulicaria* (Slobodkin and Richman, 1956), *Daphnia obtusa* (Slobodkin, 1959), and the guppy (*Lebistes reticulatus*) (Silliman and Gutsell, 1958; Silliman, 1968). None of these studies was designed to test the model of predation presented here, and thus the critical data, that is, survivorship and fecundity schedules, for testing the novel predictions of the model, such as the relationship between population size when the maximum sustainable yield is being harvested and the population's UCD, are not presented in the cited papers. Nevertheless, the results of the experiments deserve examination. In particular, the guppy study was conducted most closely to the methods employed in exploiting vertebrate populations, and its data are reported sufficiently fully to allow a comparison between its results and the model's predictions.

Silliman and Gutsell (1958) established and maintained four populations of guppies for at least 169 weeks. Two populations served as controls, and two populations were fished every 3 weeks beginning in week 40, after the populations' numbers had stabilized. Twenty-five percent of adults and immatures were removed beginning in week 40. Predation was reduced to 10% in week 79 and increased to 50% in week 121 and to 75% in week 151. Every (a) fourth, (b) tenth, (c) second, and (d) second, third, and fourth fishes, respectively, were removed every third week during the weekly censusing of the population. This method of fishing does not remove a constant proportion of each age class at each fishing effort and results in fluctuations in the populations' age structure, numbers, and yield in numbers (Fig. 5.7). The populations' biomass and yield in biomass, however, were remarkably similar for a given fishing rate (Fig. 5.8). These results indicate that these guppy populations could sustain 10 and 25% fishing rates on immature and adult age classes. With 50 and 75% fishing rates, the populations declined in numbers and biomass.

Using biomass as a measure of population size, we can see that population size decreases and yield increases with increasing harvesting rates, as suggested by the model. The populations' biomass under 25% predation is

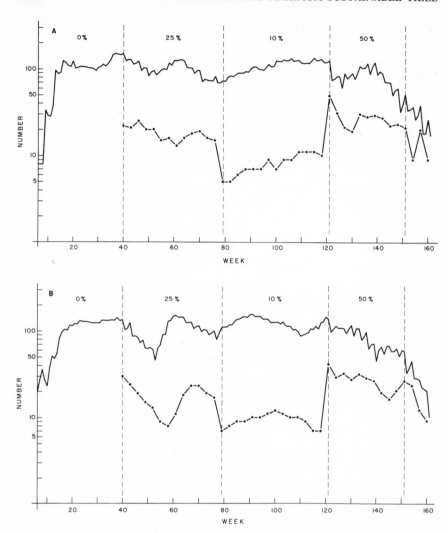

Figure 5.7. Population size and yield in numbers of two populations of guppies at different predation rates. Seventy-five percent predation, begun in week 151, drives the populations to extinction before week 175. (From Silliman and Gutsell, 1958.)

approximately at the UCD (in biomass) level (Fig. 5.8), suggesting that 25% predation may produce the maximum sustainable yield, but Silliman and Gutsell (1958) believed that the populations could sustain predation rates of 30–40%.

Later, Silliman (1968) harvested 25, 33, and 50% of guppy populations being fed different amounts of food in the ratio of 1 : 2 : 3, giving nine

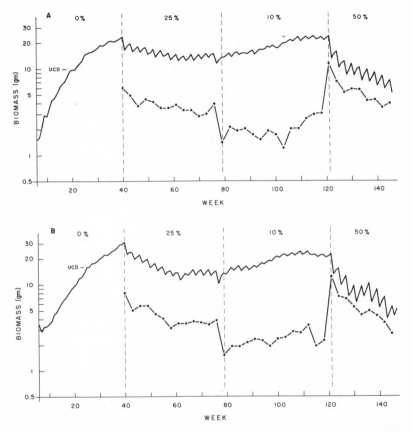

Figure 5.8. Population size and yield (measured as biomass) of two populations of guppies at different predation rates. Seventy-five percent predation, begun in week 151, drives the populations to extinction before week 175. (From Silliman and Gutsell, 1958.)

combinations. Again, biomass changes fluctuate less than numerical changes (Figs. 5.9 and 5.10). For no apparent reason, the three populations receiving the level 3 amount of food never did achieve the age distribution typical of the other populations (Fig. 5 in Silliman, 1968), and two of these populations (C and E) were marked by errors in harvesting procedures and by accidental mortality (Table 7 in Silliman, 1968), making them less useful for analysis.

Each of the nine populations began with 7 males, 8 females, and 33 juveniles (including "fry" and "immatures"). The populations were allowed to grow for 28 weeks (34 weeks in C, E, and I at food level 3) before being harvested. The greater the food supply, the larger the "steady-state" population (Figs. 5.9 and 5.10), but the ratios of the average total weights

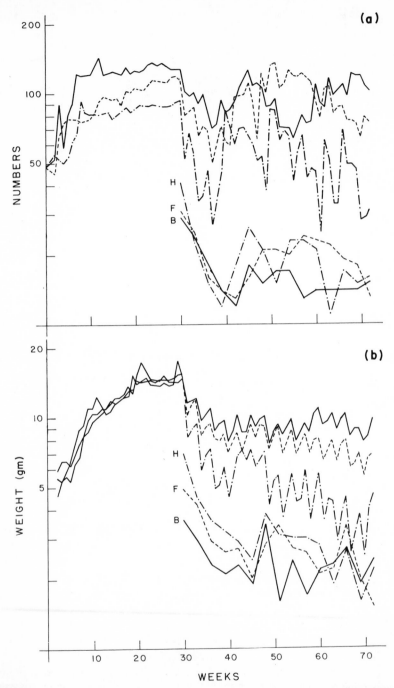

Figure 5.9. Population size and yield in numbers (a) and weight (b) for three populations harvested at different rates but receiving the same amount of food (level 1). The harvesting rates are 25% (————), 33% (-----), and 50% (-·-··-). (Data from Silliman, 1968.)

154

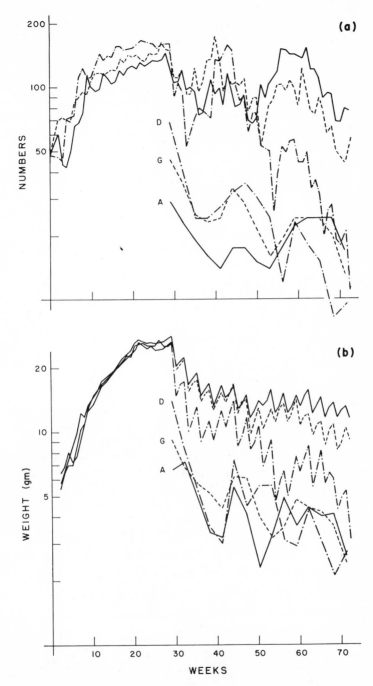

Figure 5.10. Population size and yield in numbers (a) and weight (b) for three populations harvested at different rates but receiving the same amount of food (level 2). The harvesting rates are 25% (———), 33% (-----), and 50% (-·-·-). (Data from Silliman, 1968.)

and average numbers during weeks 21–28 were 1.00 : 1.79 : 2.52 and 1.00 : 1.32 : 1.35, respectively, in comparison with the ratio of food supply, 1 : 2 : 3, indicating that some factor in addition to food was affecting the populations' growth rate. Because the fish tanks were the same size, perhaps at the higher food levels, aggression from crowding decreased the individuals' efficiency in capturing and metabolizing food or increased energy consumption.

Population sizes, in numbers and biomass, are larger with greater amounts of available food (Figs. 5.9, 5.10). Furthermore, the greater the predation rate, the greater the yield and the smaller the total population size (Figs. 5.9–5.12). Silliman (1968) concluded that yields were reasonably stable after about week 49, that 33% predation rate produced the maximum yield, and that the yield to exploitation rate was independent of the food level. The first two conclusions are questionable. Regression analysis of the data of the last 30 weeks (from Tables 6–9 in Silliman, 1968) shows that yield in numbers (10 samples for each population) showed declines in 5 of 7 populations analyzed (Fig. 5.12) and yield in biomass showed declines in all 7 populations analyzed (Fig. 5.12). Populations C and E were excluded because of the harvesting irregularities and accidental mortality. Only 4 of the 14 regressions were significant, however, 1 of 6 at 33% predation and 3 of 4 at 50% predation (Table 5.5). Regressions performed on population size in numbers and biomass were more consistent. With the exception of populations A and B (both 25% predation), all populations showed declines in both numbers and biomass during the final 30 weeks, and all of these declines, except E (25% predation) and I (33% predation) were significant (Table 5.5). The lack of significance of the declining trends in yield probably results from the small sample size, 10 samplings of yield against 30 weekly censuses of total population size. Contrary to Silliman (1968), then, most populations are declining toward the end of the experiments and the maximum sustainable yield is probably closer to 25% than to 33%. Silliman (1968), however, clearly showed that the yield of a large population cannot be increased by increasing the intensity of predation. The strategy of exploiting a prey population is independent of prey population size.

The other laboratory studies of predation (Nicholson, 1954a,b, 1958a; Watt, 1955; Slobodkin and Richman, 1956; Slobodkin, 1959) do not report data that allow evaluation of the unique aspects of the predator–prey interaction described in this chapter, for instance, the relationship between the residual population size and the population's UCD size. These papers, however, provide evidence for other intuitive or predicted ideas, already proposed by others; predation lowers population size, increases yield, and changes the age distribution, and individuals of harvested popu-

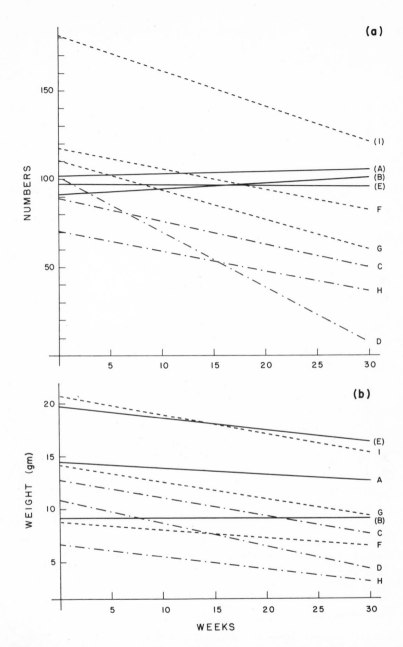

Figure 5.11. Regression analysis of population size in numbers (a) and weight (b) of nine populations of guppies harvested at different rates and receiving different amounts of food. Populations at food level 1 are B, F, and H; level 2, A, D, and G; level 3, C, E, and I. The harvesting rates are 25% (————), 33% (-----), and 50% (-·-··-). The regression analysis includes only the last 30 of 72 weeks. Nonsignificant changes in population size are indicated by parentheses at right. (Data from Silliman, 1968.)

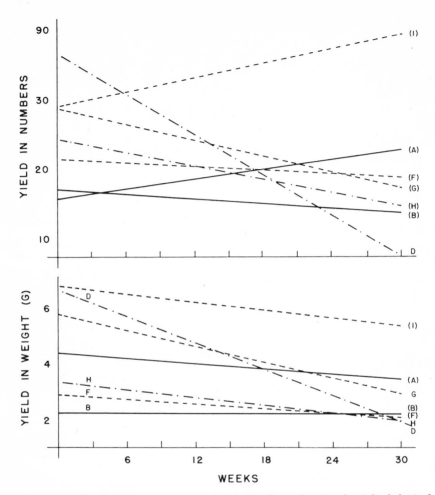

Figure 5.12. Regression analysis of yield in numbers (above) and weight (below) of seven populations of guppies harvested at different rates and receiving different amounts of food. The populations are identified in Fig. 5.11. Only the last 10 of the 14 or 15 triweekly harvests are included. Populations C and E are excluded because of errors in harvesting procedures and accidental mortality. Nonsignificant changes in yield are indicated by parentheses at right. (Data from Silliman, 1968.)

lations have greater fecundity and survivorship. Furthermore, Slobodkin (1959) harvested fixed percentages of *Daphnia* populations, taking a high proportion of immatures in some and a high proportion of adults in others. Preying on immatures lowers yield compared with preying on adults.

These experiments are important contributions to understanding several aspects of the predator–prey interaction. In order to test the means de-

TABLE 5.5

Regression Analysis on Population Data of Guppy Populations[a]

Food level:	1			2			3		
Fishing rate:	0.25	0.33	0.50	0.25	0.33	0.50	0.25	0.33	0.50
Population:	B	F	H	A	G	D	E	I	C
I. Population numbers (d.f. 1/28)									
F	0.699	10.443	16.414	0.048	12.012	128.501	0.009	3.056	4.822
p	n.s.	<0.01	<0.01	n.s.	<0.01	<0.01	n.s.	n.s.	<0.05
II. Population biomass (d.f. 1/28)									
F	0.002	20.146	40.228	4.547	26.876	40.855	2.861	13.187	12.841
p	n.s.	<0.01	<0.01	<0.05	<0.01	<0.01	n.s.	<0.01	<0.01
III. Yield numbers (d.f. 1/8)									
F	5.232	0.574	5.171	3.292	4.756	35.914	—	0.859	—
p	n.s.	n.s.	n.s.	n.s.	n.s.	<0.005	—	n.s.	—
IV. Yield biomass (d.f. 1/8)									
F	0.009	1.354	5.774	0.751	7.738	18.520	—	0.985	—
p	n.s.	n.s.	<0.05	n.s.	<0.05	<0.005	—	n.s.	—

[a] From Silliman (1968).

scribed above for determining the maximum sustainable predation rate, other experiments are required.

PRUDENT PREDATION AND MAXIMUM SUSTAINABLE YIELD

The analysis of the predator–prey interaction presented in this chapter provides some answers to questions raised in trying to understand prudent predation and maximum sustainable yield. A prudent predator, as Slobodkin (1961) has written, will surely "consume its prey in such a way as to maximize its own food supply while at the same time minimizing the possibility that the prey population will be unable to maintain itself and serve as food in the future" (p. 138). There is general agreement that a prudent predator should harvest individuals with low reproductive value (Slobodkin, 1961, 1968, 1974; Maiorana, 1976; Mertz and Wade, 1976). These belong to the older age classes, and it is shown that a maximum sustainable yield is achieved when the predator does in fact harvest the older age classes (Table 5.3). This is not quite the same as taking prey that are about to die from other causes (Slobodkin, 1961, 1968, 1974), inasmuch as many of these individuals do have several years of reproductive life left (Table 5.3). Harvesting individuals of low reproductive value, however, does not mean that yield is low (Slobodkin, 1968; Mertz and Wade, 1976). It is in fact maximal, if it is sustainable.

Not all predators have the option to be prudent. Many predators are restricted to hunting eggs, larvae, or other prereproductive individuals because of their size, behavior, or opportunities. Furthermore, young animals are often easier and safer to capture than healthy, mature animals. The prudent predation rate for a predator on younger age classes can be calculated in the manner described above, but harvesting the younger age classes does provide a smaller yield than preying on older age classes.

A predator should take whatever prey is easy to find, capture, and digest, but prudence requires restraint. There seems to be no mechanism outside of group selection to ensure the evolution of prudent restraint on the part of the individual predator, much less a predator population (Slobodkin, 1968, 1974; Maynard Smith and Slatkin, 1973; Maiorana, 1976). Slobodkin (1974) explained the apparent prudence of predators as a consequence of the evolutionary response of the prey population to the predation; that is, the prey's adaptations make it difficult for the predator to be imprudent. However, unless a predator population is limited by territorial behavior, by the time in which to breed, or by another predator, the predator population is likely to grow in size and impose an increasing intensity of predation on the prey population, eventually depressing the birth rate below the replacement rate and causing the prey population to decline. The reduction of the prey population causes the predator population to decline, as described in Chapter 3 (Fig. 3.8). The apparent prudent predators are those species whose populations are limited by space or time in which to breed or by another predator rather than by their food supply. Prudent predation seems limited to the human species, whose members are able to project and evaluate the consequences of their actions.

In order to determine the predation rate that will generate a maximum sustainable yield, the predator must know either (1) the survivorship and reproductive schedules of the prey or (2) the UCD point of the prey population's growth without predation. With the first set of information, we can calculate the predation rate that can be sustained by the prey population. This is straightforward and unambiguous. With the second, the predation rate can be regulated such that the prey population's size is maintained at the UCD level. Either set of information is difficult to establish and is presumably unavailable to the nonhuman predator.

The predator–prey interaction described in this chapter concerns the biological consequences of harvesting prey populations. The model provides a means for a prudent predator to determine the maximal predation rate that a prey population can withstand and still provide a maximum sustainable yield to the predator. Implementing an effort that provides only a maximum sustainable yield often requires restraint, especially for the human predator who has developed technologies that enable him to pursue profitably even rare species. The economics of exploiting the commons or open-access fisheries, however, encourages overharvesting rather than restraint (Hardin, 1968; Clark, 1973, 1976). As long as profits accrue to the individual and costs are spread among everyone, and as long as the return on investing profits of present overharvesting exceeds the expected returns on future harvests, the overexploitation of resources is inevitable. Certainly, if whalers, for example, still had to harpoon whales by hand from small boats, the whale population would be larger than it is, and, if everyone benefited from the existence of whales, societies or nations would institute conservation measures and proscriptions against imprudent individuals or nations.

CONTROL OF PEST POPULATIONS

The primary stimulus for thinking about population dynamics has been the economic interest in controlling the depredations of insect pests. Entomologists have led the way in developing population theory (Howard and Fiske, 1912; Nicholson, 1933, 1954a,b, 1958a,b; Smith, 1935; Thompson, 1939, 1956; Solomon, 1949, 1957, 1964; Andrewartha and Birch, 1954, 1960; Andrewartha, 1958; Birch, 1958; Milne, 1957, 1958a, 1962; Huffaker, 1958a; Schwerdtfeger, 1958; Huffaker and Messenger, 1964a; Clark *et al.*, 1967) in addition to their dealing with purely practical matters.

The two techniques used in controlling insect populations are biological and chemical—the introduction of predators or pathogens and the application of chemical pesticides. Despite the differences in the purposes of managing a prey population in order to provide a maximum sustainable yield and of controlling the size of pest populations in order to reduce their depredations on economically important crops, the dynamics of managed populations is the same.

Predators and pathogens increase the pest population's mortality rate and raise its replacement rate in the manner already described (Figs. 3.8 and 5.3), and the pest population becomes predator- or pathogen-limited rather than food-limited. Predators and pathogens differ in their life history characteristics and in their abilities to find, capture or infect, and kill or debilitate the prey or host. Each predator or pathogen will have different effects on the prey population (Fig. 5.13). Whereas the manager attempting

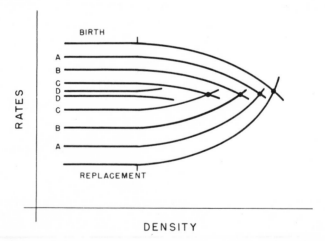

Figure 5.13. Effects of biological and chemical controls on pest populations. Predators or pesticides with different effects on the pest population's birth and replacement rates are A, B, C, and D. More effective predators or pesticides reduce both the pest's growth rate and "steady-state" size more greatly than do less effective predators or pesticides. Predator or pesticide D will eventually eliminate the pest population, but C, B, or even A may effectively reduce the pest's depredations because with the lower growth rate the pest does not have sufficient time to reach a size that can cause damage.

to maintain a maximum sustainable population will not permit the predation rate to reach a level that raises the prey's replacement rate above its birth rate, the manager of a pest population will search for the predator or pathogen whose use is practical and economic in reducing the pest population to a desired size. It is not necessary to annihilate the pest population or even to raise its replacement rate above its birth rate. For example, if pest population N_C (Fig. 5.13) does not cause economic damage, there is no point in searching for or applying a more effective predator or pathogen. Furthermore, predators and pathogens always reduce the pest population's growth rate (r), even below the UCD (Figs. 5.3 and 5.13). Thus, predators and pathogens may change the food-limited pest population to a time-limited population (Fig. 3.13). Because of its lower growth rate, the potential pest does not have the time during its breeding season to reach a size of economic importance.

The preeminently successful case was the introduction of the cactus moth (*Cactoblastis cactorum*) into Australia to control *Opuntia* sp. (Dodd, 1959), described earlier (pp. 8–9). This and other cases are often reviewed (e.g., Andrewartha and Birch, 1954; Clark *et al.*, 1967).

Different pesticides or different concentrations of the same pesticide also will have different effects on the pest population's birth and replacement rates (Fig. 5.13). The considerations involved in selecting a particular predator or pathogen are applicable in selecting the pesticide and its concentration to control a pest population.

6 | COMPETITION

The central proposition of competition theory is the competitive exclusion principle, which may be identified as the third law of population dynamics: *competing populations cannot coexist indefinitely*. Interspecific competition occurs whenever the individuals of two or more species are using the same resources that are in short supply (Milne, 1961). Andrewartha and Birch (1954) and Birch (1957, in his meaning I) add that, if the common resource is not in short supply, interspecific competition occurs when the individuals harm one another in acquiring that resource. To be explicit, harm refers to a decline in fecundity, to an increase in mortality, or both. The consequences of interspecific competition are that either one or both species will adapt and use the environment's resources in different ways or one of the populations will become extinct.

It is easier to define "resources in short supply" for a theoretical discussion than it is to demonstrate a short supply in nature. What may appear to be an abundant resource to the human observer may not in fact be abundant to the animals consuming it. Individual items of a resource can vary in their availability and quality, and such variation may not be readily apparent to the investigator. For example, the leaf-eating howler monkey (*Alouatta palliata*) is limited to plant parts low in alkaloids, and leaves vary considerably in alkaloid concentration from tree to tree of the same species (Glander, 1977). Thus, the abundance of suitable leaves for howler monkeys is less than what it appears to be. Nevertheless, whatever the problems are in assessing the quantity of resources in the field, for theoretical purposes the assumption that interspecific competition occurs whenever the probabilities of survival or successful reproduction are lowered seems reasonable.

Darwin (1872) certainly recognized the consequences of interspecific competition, but it was the competition equations of Volterra (1931) and Lotka (1932), the experimental studies of Gause (1934), and the many cases of differences in habitat, body size, and foraging behavior of sympatric populations (Lack, 1944, 1954, 1971) that established the competitive exclusion principle as an important ecological generalization.

The competition equations of Volterra and Lotka can be represented as follows:

$$\frac{dN_1}{dt} = r_1 N_1 \left(\frac{K_1 - N_1 - \alpha N_2}{K_1} \right),$$

$$\frac{dN_2}{dt} = r_2 N_2 \left(\frac{K_2 - N_2 - \beta N_1}{K_2} \right),$$

where N_1 and N_2 are the population sizes of species 1 and 2, respectively; r_1 and r_2 are the growth rates of the two species; K_1 and K_2 are the saturation values of each species when grown alone; α and β are the competition coefficients.

Mathematical and graphical analyses (Lotka, 1932; Gause and Witt, 1935; Andrewartha and Birch, 1954; Slobodkin, 1961) show four solutions to these equations (Fig. 6.1); (I) species 1 is always the winner; (II) species 2 is always the winner; (III) either species 1 or species 2 will be the winner depending on their initial densities; and (IV) species 1 and 2 coexist. Case (IV) is often described in textbooks as representing a stable equilibrium between species whose intraspecific competition is more intense than their interspecific competition. Gause and Witt (1935), however, originally described it as a "stable combination, in which each of the species is driven into its 'niche'" (p. 600) within a heterogeneous habitat with each species using resources that are not available to the other species. Thus, case (IV) represents the consequences of competition in heterogeneous habitats, and the other cases represent the consequences of competition in homogeneous habitats, at least with respect to the factors limiting population growth. Interspecific competition always leads to exclusion.

Andrewartha and Birch (1954) strongly criticized the Lotka–Volterra competition equations, and Cole (1960) admonished ecologists to apply the competitive exclusion principle with caution. Ayala (1969) presented data

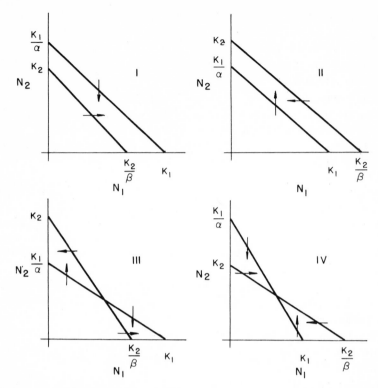

Figure 6.1. Competitive interactions between two species, according to the Lotka–Volterra equations. Here, N_1 and N_2 are the sizes of populations 1 and 2, respectively. The line K_1, K_1/α represents the maximal size of species 1 at various densities of species 2, and the line K_2, K_2/β represents the maximal size of species 2 at various densities of species 1. The horizontal arrows indicate the direction in which the population size of species 1 grows (decreases to the left) at the maximal size of species 2. The vertical arrows indicate the direction in which the population size of species 2 grows (decreases toward the bottom) at the maximal size of species 1. There are four possible consequences: (I) species 1 is always the winner; (II) species 2 is always the winner; (III) either species 1 or species 2 will be the winner, depending on their initial densities and growth rates; and (IV) species 1 and species 2 "coexist." "Coexistence" occurs in heterogeneous habitats in which each species uses resources not available to the other. (Redrawn from Gause and Witt, 1935.)

that he interpreted as constituting evidence for rejecting the principle. Nevertheless, it remains one of the central principles of ecology (Hardin, 1960; Miller, 1967; Lack, 1971). It is included in virtually every ecology textbook, usually without qualification.

The interpretation of population dynamics being developed in this book has rejected the Verhulst–Pearl logistic equation as a valid model of population growth. Because the Lotka–Volterra model of interspecific competition is based on the Verhulst–Pearl equation, it must also be rejected as a valid model of the dynamics of interacting populations, a view already proposed by Andrewartha and Birch (1954).

Although the Lotka–Volterra model must be rejected, the competitive exclusion principle need not be rejected. Intuitively, the principle is appealing, and many data from both field and laboratory are consistent with it. Sympatric species almost always differ in their habitat, foraging behavior, or size (examples reviewed in Lack, 1944, 1954, 1971; Crombie, 1947; and Miller, 1967). This is especially striking when two species of similar habitat, foraging behavior, or size where allopatric differ in habitat, foraging behavior, or size where sympatric (reviews in Brown and Wilson, 1956; Mayr, 1963; Miller, 1967; and Lack, 1971), a phenomenon called character displacement (Brown and Wilson, 1956). Laboratory studies, beginning with Gause (1934), have tended to support the competitive exclusion principle (Miller, 1967). Criticisms of the competitive exclusion principle and difficulties in interpretation are discussed below. Let us now establish a theoretical basis and the proper meaning for the principle.

Consider two species, A and B, which are food-limited. Species A is better able to use the resource(s), and when grown alone reaches a greater population size (or biomass) than does B when it is grown alone (Fig. 6.2). When grown together, A has a greater impact on B than vice versa, depressing B's survivorship and fecundity more than B depresses A's. The relative inefficiency of B allows A to grow, even at high densities of B (Figs. 6.2 and 6.4a). Species A will grow, and as it increases in density it depresses the growth rate of B until it is less than zero, and B becomes extinct. This must always be so because A grows at any density of B. Although B's presence causes A to grow more slowly, B cannot prevent A from attaining its full potential. Therefore, species B cannot coexist with A.

There is the possibility that at high densities species B can depress A's birth rate below its replacement rate so that A's population declines to extinction (Figs. 6.3 and 6.4b). In this case, either species can be the winner. The population that first attains the size that depresses the size of the other, because of a difference in intial population density or a difference in growth rate, will be the winner.

It might be tempting to suggest that even at maximum density neither species can depress the other's birth rate below the replacement rate (Fig. 6.4c), and thus the coexistence of competitors occurs. As with case (IV) in Fig. 6.1, however, this condition indicates that there are resources in the system that are not used by one of the species in sustaining its own growth, that is, the system is heterogeneous. The assumption of the competitive exclusion principle, however, is that both species are using resources in

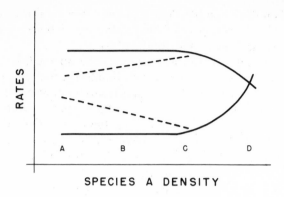

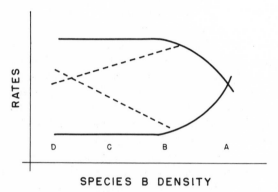

Figure 6.2. Birth and replacement rates of two competing species. Solid lines show the rates in the absence of competition; the dashed lines show the rates in the presence of the competitor. These rates vary with the relative densities of the competitors, as shown at A, B, C, and D. Species A is the better competitor. As its population grows, it depresses further the growth rate of species B. Species A is always the winner because it can grow at any density of B (see also Fig. 6.4).

common. When a species reaches its maximum density, there should be no resource remaining for the use of the second species.

The competitive exclusion principle is often misinterpreted or misapplied. A plethora of papers begins with the assumption that coexisting species cannot be competing and ends by demonstrating some difference between the two species, which is supposed to account for the lack of competition and for coexistence. The authors have followed the lead of Lack (1944, 1945, 1946, 1947b, 1954, 1966, 1971), who consistently stated (e.g., 1947b, p. 62), "Two species with similar ecology cannot live in the same region." Even Hardin (1960) has written, "Complete competitors

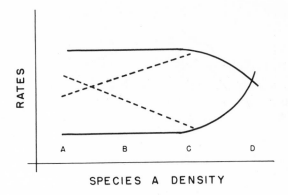

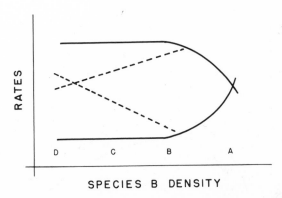

Figure 6.3. Birth and replacement rates of two competing species. These figures are similar to those in Fig. 6.2, except that high densities of species B does depress species A's growth rate below zero. In this competitive relationship either species can win. The winner is determined by which species first reaches the density that depresses the other species' growth rate below zero (see also Fig. 6.4b).

cannot coexist" (p. 1292), when it is clear that he meant otherwise. These statements and others cited by Gilbert *et al.* (1952) are misleading because they imply that coexisting species must differ ecologically in some way.

The competitive exclusion principle predicts the *consequences* of interspecific competition. However, for interspecific competition to occur, the competing species *must* coexist for a time, even if they cannot coexist indefinitely. Such coexistence could be prolonged depending on the degree of competition. Thus, coexistence itself is not sufficient to demonstrate a lack of competition. Even ecological differences do not demonstrate a lack of competition, unless it can be shown that the differences are related to the use of a limiting resource. For example, if food is not limiting the

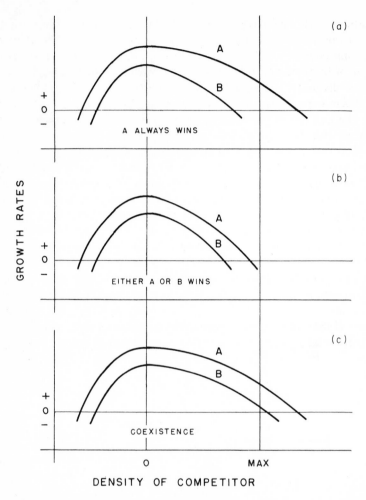

Figure 6.4. Growth rates of competing species. (a) Only species A can win because only it has a positive growth rate at all densities of its competitor. (b) Either species can win because neither species has a positive growth rate at all densities of its competitor. (c) This relationship results in coexistence because each species has a positive growth rate at all densities of its competitor. Because each species can grow when the other species is at its maximal density, each species must be utilizing some resource that is unavailable to the other species.

populations of two sympatric species, the differences in feeding structures and behavior between the two species cannot be assumed to be adaptations that permit coexistence. Despite the difficulties involved in studying competition, within the voluminous literature there seem to be a few cases, among a diversity of taxonomic groups, which are best interpreted as examples of competitive exclusion (Crombie, 1947; Miller, 1967). In these cases, the habitat use of each of the sympatric populations is restricted compared with their use of habitats where the populations are allopatric.

An early example was described by Beauchamp and Ullyott (1932). They showed that two species of freshwater planarians (*Planaria montenegrina* and *P. gonocephala*) in southern Europe occupied a wider range of temperature in streams when alone than when both species occurred in the same stream (Fig. 6.5). When alone, *P. montenegrina* was found at temperatures from 6.6° to 17°C, and *P. gonocephala* was found at temperatures from 8.5° to 23°C. When both species occurred in the same stream, only *P. montenegrina* was found at temperatures below 13°–14°C, and only *P. gonocephala* above 13°–14°C. Beauchamp and Ullyott (1932) interpreted this pattern of distribution to be evidence for the existence of competition between these two species.

Gause (1934) undertook a series of experiments with protozoans and yeast cells in laboratory cultures. In a competition experiment, he compared the growth of *Paramecium aurelia* and *P. caudatum* when grown alone and

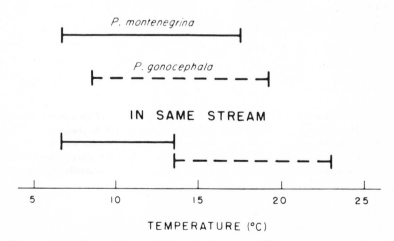

Figure 6.5. Competitive exclusion in flatworms, *Planaria montenegrina* and *P. gonocephala*. Each species occupies a wider range of temperatures when found in different streams than when found in the same stream. (Data from Beauchamp and Ullyott, 1932.)

when grown together in cultures provided with a constant quantity of bacteria as food. When alone, *P. aurelia* grew faster and reached a greater volume than did *P. caudatum*, but growth in both cases was typically sigmoid (Fig. 6.6). In cultures provided with twice as much food, the protozoan populations approximately doubled in size, showing that the amount of food was the factor limiting the population size of these species. When grown together, both species increased until the fifth day, after which *P. aurelia* increased at the expense of further growth of *P. caudatum*, which virtually disappeared from the cultures (Fig. 6.6).

Connell (1961) studied competition in two barnacle species in Scotland. Adult *Chthamalus stellatus* occupy the rocky shoreline between the levels of the mean high water of neap and spring tides, and adult *Balanus balanoides* occupy the shoreline at lower levels, between the high and low water of neap tides (Fig. 6.7). The larvae, however, settle throughout a much wider range of heights (Fig. 6.7). The upper boundary of *Balanus* distribution is determined by high mortality at the higher, drier levels. The lower boundary of *Chthamalus* distribution, however, is determined by the presence of *Balanus*, which undercut or crush the *Chthamalus* individuals that are present in the lower intertidal range. That *Chthamalus* could survive at lower levels was shown experimentally by removing *Balanus* from selected areas. This is a clear case of the distributions of two species being determined by interspecific competition for space.

Interspecific competition between space-limited bird populations takes the form of interspecific territoriality (Murray, 1969, 1971, 1976). The species whose members have a better than 50% chance of winning interspecific aggressive encounters and establishing territories will eventually exclude the other species from their common habitat. Cases of mutual interspecific territoriality in a common habitat should be rare, and in fact they are. Mutual interspecific territoriality occurs at the edges of two species' ranges, as between the Red-bellied Woodpecker (*Centurus carolinus*) and Golden-fronted Woodpecker (*C. aurifrons*) in Texas (Selander and Giller, 1959), the Eastern Meadowlark (*Sturnella magna*) and Western Meadowlark (*S. neglecta*) throughout the zone of recent overlap in the central United States (Lanyon, 1956, 1957), and the Melodious Warbler (*Hippolais polyglotta*) and Icterine Warbler (*H. icterina*) in western Europe (Ferry and Deschaintre, 1974), or at the edges of distinctly different habitats, as between the Dusky Flycatcher (*Empidonax oberholseri*) and Gray Flycatcher (*E. wrightii*) in the western United States (Johnson, 1963, 1966) and the Black-winged Red Bishop (*Euplectes hordeacea*) and Zanzibar Red Biship (*E. nigroventris*) in eastern Africa (Fuggles-Couchman, 1943).

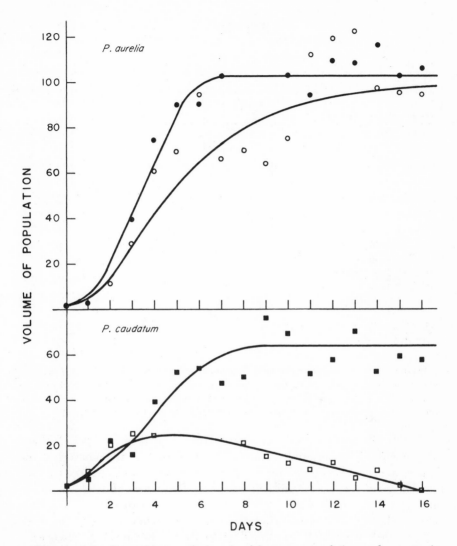

Figure. 6.6. Competitive exclusion in laboratory populations of paramecia, *Paramecium aurelia* and *P. caudatum*. The closed symbols represent counts when each species was grown alone. The open symbols represent counts from mixed cultures. (Data from Gause, 1934.)

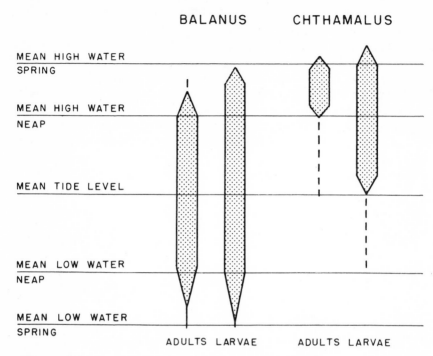

Figure 6.7. Competitive exclusion in populations of barnacles, *Chthamalus stellatus* and *Balanus balanoides*. The intertidal ranges of newly settled larvae are greater than those of adults. The upper level of *Balanus* is determined by desiccation, but the lower level of *Chthamalus* is determined by the presence of *Balanus*. (From Connell, 1961.)

Widely sympatric species occupying the same habitats either behave differently or are different in size. The Red-winged Blackbird and the Tricolored Blackbird (*Agelaius tricolor*) occupy marshes in central California (Orians, 1961; Orians and Collier, 1963). Red-winged Blackbirds establish large territories. When the nomadic Tricolored Blackbirds arrive on the marsh, they establish small territories, several in each Red-wing territory, and when a Tricolor is chased by the established territorial Red-wing it simply flies away and returns when the Red-wing chases another Tricolor. The Red-winged Blackbird tires, and the Tricolors breed successfully. On other marshes in the western United States, Red-winged Blackbirds are involved in interspecific encounters with the Yellow-headed Blackbird (*Xanthocephalus xanthocephalus*) (Orians and Willson, 1964). In this case, the larger Yellow-headed Blackbirds simply evict the Red-wings from their territories. Yellow-headed Blackbirds nest in vegetation in deeper water, and interspecific aggression persists at the boundaries between the Yellow-headed Blackbirds and the Red-winged Blackbirds in adjacent shallower water.

Interspecific territoriality will result not only in the evolution of different territorial behaviors or different size but also in the evolution of habitat differences (Fig. 6.8). The latter will be difficult to demonstrate but not more difficult than demonstrating that present habitat differences result from some other form of competition having occurred in the past. These expectations are consistent with the competitive exclusion principle.

Perhaps the most convincing evidence supporting the competitive exclusion principle has been the phenomenon called character displacement, in which the sympatric populations of two species are more different in

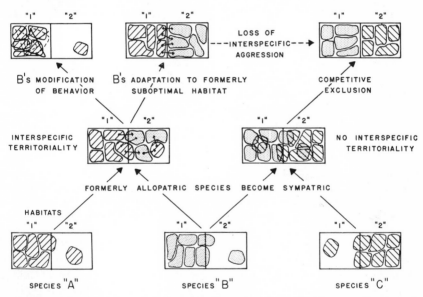

Figure 6.8. Consequences of interspecific territoriality and of interspecific competition. Species A and species B are best adapted to habitat 1, while species C is best adapted to habitat 2. If species B and C become sympatric, compete for some resource, and are not interspecifically territorial, their co-occupancy of habitats 1 and 2 (center right) would not persist indefinitely before competitive exclusion resulted in habitat segregation (upper right). If species A and B become sympatric, do not compete for resources (e.g., food, nesting sites) other than territorial space, and are interspecifically territorial, then species B, if subordinate to A, will be forced out of its optimal habitat (center left). Species B may either modify its territorial behavior (upper left) or become adapted to its suboptimal habitat (upper center). If species B subsequently loses its interspecific territoriality through divergence resulting from intraspecific selection for different color pattern or vocalizations, then habitat segregation not distinguishably different from that resulting from competitive exclusion could occur. (From Murray, 1971.)

structure and behavior than their allopatric populations (Brown and Wilson, 1956; Hutchinson, 1959; Mayr, 1963; Miller, 1967). Character displacement is a predictable consequence of the competitive exclusion principle. If the ranges of two similar species overlap, and if competition occurred in the past, we should expect the species to differ in one way or another where sympatric, as a consequence of selection favoring the reduction of competition, but to remain similar in allopatry where competition and therefore selection for divergence do not occur.

The most frequently cited case of character displacement is that of the Rock Nuthatches (*Sitta neumayer* and *S. tephronota*), described originally by Vaurie (1950, 1951). The ranges of the two species overlap in the Middle East (Fig. 6.9). To the east and west of the overlap zone the average bill lengths of *S. tephronota* and *S. neumayer*, respectively, are about the same size (Fig. 6.9). Where sympatric, however, the bills of *S. tephronota* are larger and those of *S. neumayer* are smaller than where each species

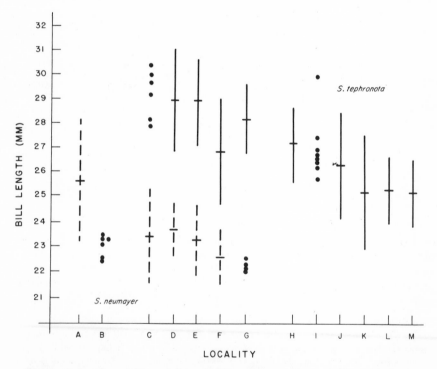

Figure 6.9. Geographic variation of bill length in two species of Rock Nuthatches, *Sitta neumayer* and *S. tephronota*, from Greece (population A) eastward to Tian Shan (population M). The zone of overlap is in Iran (populations C–G), where bill lengths do not overlap. (From Vaurie, 1951.)

occurs alone, indicating that divergence of bill length occurred in the zone of overlap. This is exactly what the competitive exclusion principle predicts.

Grant (1972), however, has reexamined this and the other cases of character displacement in birds and suggested alternative explanations for the observations. In the case of the nuthatches, the bill lengths of each species vary in a cline from west to east, each species being larger in the west. The westernmost populations of *S. tephronota*, which have the larger bills, overlap with the easternmost populations of *S. neumayer*, which have the smaller bills. Thus, the apparent divergence in the area of sympatry appears to be a consequence of each species responding similarly to its physical environment. Which explanation is better requires more information.

A further complication in interpreting character displacement is that in some species of *Sitta* (Ripley, 1959), in *Centropus* (Parkes, 1965), and in other cases (Brown and Wilson, 1956) character displacement involves the coloration and pattern of plumage, differences that may have evolved as premating isolating mechanisms permitting specific recognition. Perhaps character differences that seem to have ecological significance, such as bill size differences in birds, also evolved as premating isolating mechanisms, as suggested by Vaurie (1951), Brown and Wilson (1956), and Miller (1967). Indeed, Lack (1947b) presented evidence that the bills of some species of Galapagos Finches (Geospizidae) do serve as specific recognition marks, but he concluded that their bill differences were adaptations that eliminate competition for food. Even though he did not demonstrate that food was limiting, he used the competitive exclusion principle as justification for his interpretation.

An issue more important than the quality of the evidence for or against the competitive exclusion principle is whether it can be tested at all (Gilbert *et al.*, 1952; Hardin, 1960; Cole, 1960). If extinction does not occur, a proponent of the principle can conclude that the two species differ in some way that had not been detected. With few exceptions (Ross, 1957), further effort can turn up differences between species, and numerous papers demonstrate some minor difference between coexisting species without attempting to demonstrate a relationship between the difference and the potential competition.

Hardin (1960) correctly pointed out that the competitive exclusion principle is one of those scientific generalizations that cannot be tested directly. Scientists, for instance, did not reject Galileo's formulation for the distance traveled by a free-falling object simply because a feather and a hammer on earth fell different distances in the same time. All generalizations are applicable to a given set of restrictive conditions, and Galileo's theory was tested directly only after men were able to land on the moon, where conditions suitable for the direct test were available.

The competitive exclusion principle is applicable and can be tested directly only when two (or more) populations are using a resource that is in short supply, all other factors being held constant. Competitors can coexist (perhaps indefinitely) when the other factors are not held constant.

1. If the environmental conditions are changing such that each species is favored alternately, the two species could coexist even though competition between them is occurring constantly (Crombie, 1947; Hutchinson, 1948). An example is the coexistence of *Drosophila melanogaster* and *D. funebris* in laboratory cultures (Merrell, 1951). *Drosophila melanogaster* did better on fresh food, whereas *D. funebris* did better on older food. The periodic introduction of fresh food, which then aged, led to fluctuations in food quality, and the two populations coexisted for two years, after which the experiments were terminated. In another experiment, when fresh food was not added, only *D. funebris* survived.

In another laboratory study (Utida, 1957), two species of wasps (*Neocatoloccus mamezophagus* and *Heterospilus prosopidus*) parasitizing the larvae of a single species of beetle (*Callosobruchus chinensis*) coexisted for about 70 generations over a 4-year period. The numbers of all three populations varied greatly during this time. *Heterospilus* was the more efficient parasite when the beetle population was low, and *Neocatoloccus* was more successful at high beetle densities. Thus, the competition coefficients changed with varying host densities, and the competing species were able to coexist (perhaps) indefinitely.

These examples of coexisting competitors do not constitute evidence for rejecting the competitive exclusion principle.

2. Cole (1960) drew attention to Skellam's (1951) interesting statistical models showing how competing species could coexist. If the environment contained a given number of spots on which one individual of either species A or species B could grow and produce seed, which disperse randomly, both species could coexist in the area if (a) each species wins 50% of the time on the spots seeded by both species (an unlikely possibility) or (b) the poorer competitor (which loses more than 50% of the time) produces more seeds than the better competitor. Hutchinson (1958) suggested that this model had limited application (to annual plants with a definite breeding season), but Cole (1960) is probably correct in extending its application to other systems.

The interpretation of these models, however, is affected by the size of the area being considered, as are other aspects of population biology. In the larger area the two competitors are certainly coexisting, but in the smaller spots they are just as certainly not coexisting. In fact, the coexistence of the two species in the larger area is irrelevant with respect to the further evolution of the species. Those seeds that land on spots occupied by their competitor and therefore do not develop are selected against. Even though seeds of the poorer competitor do well when they fall on spots unoccupied by the better competitor's seeds, there will be selection for any changes that lead to successful germination and development on spots occupied by the competitor. The eventual evolution of differences between the species is expected. Although ecologists and evolutionists focus attention on populations, natural selection selects individuals.

3. Hutchinson (1948) and Slobodkin (1961) showed that if a predator or other environmental factor reduces the populations of two competing species, the competitors can coexist, sometimes indefinitely. Such predator-limited populations, however, are not competing because the resource they are using in common is not a limiting factor to either population, and thus the competitive exclusion principle is not applicable.

For example, Paine (1966) removed the predaceous starfish *Pisaster ochraceus* from an intertidal marine invertebrate community of 15 species in the northwestern United States. Within 3 months the barnacle *Balanus glandula* occupied from 60 to 80% of the available space, but within 1 year *Balanus* was being crowded out by the mussel *Mytilus californianus* and the goose-necked barnacle *Mitella polymerus*. Species richness dropped from 15 to 8. Evidently, *Pisaster* predation limited the growth of the dominant competitor populations (*Balanus*, *Mitella*, and *Mytilus*) and prevented them from occupying all the space on the rocky substrate, allowing opportunities for the other species (e.g., limpets, chitons) to survive.

4. Despite the inherent difficulties in developing tests for the competitive exclusion principle, Ayala (1969) rejected the principle on the basis of data that he claimed show the coexistence of two competing species. Ayala grew *Drosophila serrata* with *D. pseudoobscura* AR and with *D. pseudoobscura* CH for 40 weeks. In all 11 experimental populations, *D. serrata* declined in frequency during the last 25 weeks or so with *D. pseudoobscura* AR and during the last 15 weeks or so with *D. pseudoobscura* CH (Fig. 6.10). Had the experiments continued, it seems possible that *D. serrata* could have been eliminated. Curiously, Ayala (1969) adds, "There is direct evidence that *D. serrata* is superior in competitive fitness to *D. pseudoobscura* at the larval stage but inferior to it at the adult stage" (p. 1079). If so, the competition coefficients are changing with the stage of the life cycle, which could lead to prolonged, if not indefinite, coexistence without violating the competitive exclusion principle. If the superiority of *D. serrata* in the larval stage is not equal to the superiority of *D. pseudoobscura* in the adult stage, sooner or later we should expect exclusion.

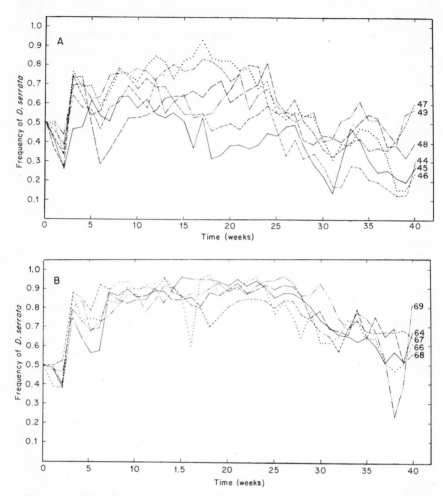

Figure 6.10. Results of competition between *Drosophila serrata* and *D. pseudo-obscura* AR (A) and between *D. serrata* and *D. pseudoobscura* CH (B). (From Ayala, 1969.)

To summarize, as with other scientific principles, the competitive exclusion principle is a useful generalization that allows us to think about and to analyze what is happening in natural populations. It does not tell us what is happening. We expect populations to adapt to the presence of predators, disease organisms, and pesticides, factors that depress survival and fecundity. Surely, we should expect populations to adapt to the presence of competitors, which also depress survival and fecundity.

There is no justification for assuming that coexisting populations are not competing or for interpreting every difference between sympatric populations as adaptations to the presence of competitors or to past competition. Coexisting species with quite different structures and behaviors may very well be competing.

The problem of interspecific competition for ecologists resembles the problem of speciation for systematists. First, with respect to speciation, we assume that two populations that coexist without interbreeding possess premating isolating mechanisms. Second, if some individuals of the two populations interbreed and produce inferior or no offspring, we expect premating isolating mechanisms to evolve. Third, we do not often know whether existing premating isolating mechanisms are incidental consequences of divergent evolution of formerly allopatric populations or are the result of selection because of the disadvantages of hybridization.

In exactly parallel fashion, first, many ecologists assume that ecological differentiation is the necessary condition for indefinite coexistence. Second, if two populations come together and compete for some resource, we expect differences to evolve. Third, we usually have no idea whether existing ecological differentiation is an incidental consequence of divergence having occurred between formerly allopatric populations or is a consequence of character displacement, resulting from the disadvantages of interspecific competition.

The occurrence of hybridization does not lead systematists to reject their theory of speciation, and the coexistence of competing species should not lead ecologists to reject the competitive exclusion principle. The critics of competition and the competitive exclusion principle (Andrewartha and Birch, 1954; Cole, 1960) are not calling for the rejection of competition as a valid interpretation of certain observations, but they are objecting to the widespread misapplication of the term.

COMPETITION WITHOUT SHORTAGES

Brown and Wilson (1956) and DeBach (1966) suggested that a limited supply of resources is an unnecessary condition for the occurrence of competition. In their view, competition occurs whenever two or more species are using a common resource. In a further elaboration, Wilson (1975) recognized degrees of competition. Competition for unlimited resources has no ecological consequences, but when resources are limited competition results in ecological exclusion. The notion that competition occurs whenever two or more individuals or populations are using any resource in common regardless of its availability can only lead to semantic confusion and, indeed, is contrary to its dictionary definition (Urdang, 1968). DeBach (1966), however, suggested that the population with the greater growth rate will be the winner in competition for unlimited resources. He drew an analogy with natural selection: "This usage of competition fits most closely the ideas of evolutionists and geneticists concerning competition in natural selection. Dobzhansky. . . . explains that natural selection may take place when resources are not limiting, if the carriers of some genes possess greater reproductive potentials than others. 'Fitness' is merely a measure of reproductive proficiency. This applies as well to competitive displacement" (p. 192). This is not so. Competition and natural selection are not equivalent terms (Andrewartha and Birch, 1954; Birch, 1957; Milne, 1961; Williams, 1966a). If food is abundant, the individuals that are able to survive longer and raise more progeny because they are better at finding, capturing, transporting, or metabolizing food than other individuals will be selected for. Natural selection can occur in the absence of competition. Competition, however, as defined above, is a potent evolutionary force.

7 THE DYNAMICS OF POPULATIONS

There are four themes that run through the conventional wisdom of population dynamics, which I believe affect current thinking. First, ecologists follow Darwin (1872), who wrote, "There is no exception to the rule that every organic being naturally increases at so high a rate, that, if not destroyed, the earth would soon be covered by the progeny of a single pair" (p. 54). Darwin went on to calculate the geometric growth of several hypothetical populations and cited several examples of rapidly growing populations in nature to support his contention. This way of looking at the consequences of individual fecundity is important from the point of view of evolution because natural selection is concerned with which individuals of the many who are born actually survive and reproduce.

Yet dying is more certain than giving birth. Every population loses individuals to predation, disease, the rigors of the physical environment, and accidents. From the point of view of population dynamics, it could be argued that, if the surviving individuals did not produce as many offspring as they do, the population would become extinct (Fisher, 1930). The belief in the potential for rapid growth leads ecologists to think about subtractive processes and to ideas about "controlling," "governing," or "regulating" the surplus, when in fact the "struggle for existence" against the elements may be so severe that high fecundity is essential for survival.

Second, ecologists' thinking is also influenced by the kinds of populations they study. Almost all ecologists study populations whose members are sufficiently numerous to ensure developing statistically significant data in a limited period of time. Some ecologists are concerned specifically with rapidly growing populations of economically important pest species, especially those ecologists who have contributed so much to our present understanding of population dynamics. Virtually nothing is known about the multitude of less common species although, with the recent public interest in environmentalism, studies of endangered species have been undertaken. We are, however, as ignorant of why one species is common and another rare as when Darwin (1872) first commented on the relative abundance of species. Furthermore, the estimate by Simpson (1952) that some 500,000,000 species have existed on earth implies that extinction is a common occurrence. The extinction of species requires the extinction of many more local populations. Not only the potential for rapid growth but the potential for extinction must be part of our understanding of population dynamics.

Third, ecologists have fitted their theories and equations to series of population counts and focused attention on growth rates. The population's size is thought to be a consequence of variations in the growth rate, which are induced by density-dependent factors. Rather than assume that populations have evolved characteristic "intrinsic rates of increase," I propose that the dynamics of population growth can best be understood by looking at the birth and death rates that prevail under particular conditions. Just as physicists successfully solved the problem of projectile motion by considering separately the vertical and horizontal components of the forces acting on a projectile, ecologists may better explain a population's fluctuations in numbers by considering separately the birth rate and death rate components of population growth.

Fourth, ecologists tend to interpret ecological systems as equilibrium systems, whether of populations (e.g., the Verhulst–Pearl logistic equation), communities (MacArthur, 1955; Leigh, 1965; MacArthur and Wilson, 1967), or whole ecosystems (Margalef, 1968). There is little evidence that ecological systems are equilibrium systems, however. Any series of counts of individuals over a period of time in either natural or laboratory populations shows considerable variation. The only well-documented equilibrium systems in biology are the homeostatic mechanisms of individual organisms. This is exactly what one expects if natural selection is the cause of evolutionary change. Individuals with superior homeostatic mechanisms would be selected for, if the maintenance of equilibrium states is beneficial to their possessors. The evolution of adaptations of populations or higher levels of organization requires a mechanism such as group selection, as proposed by Wynne-Edwards (1962).

The view of population dynamics presented here emphasizes the effects of space, food, predators, and time on individual survivorship and fecundity in the determination of a population's size. Both survivorship and fecundity are influenced by many factors, not the least of which is natural selection, which selects for increases in individual survival and reproductive success. Nevertheless, there is no single set of survivorship and reproductive values that characterize a population. Although the variation in individual fecundity with changing conditions is known to some extent (Lack, 1954, 1966), life table constructions (Deevey, 1947; Spinage, 1972) tend to convey the idea that survivorship is characteristic of a species (or population) and is static. However, survivorship will vary with prevailing conditions of weather, predation, disease, intra- and interspecific competition, and availability of resources. The probabilities of survival may change as rapidly as environmental temperature, changing the stable age structure and replacement rate as quickly. These short-term changes in age-specific death rates are difficult to detect and to measure in the field, but we can be sure they are occurring. Survivorship has been shown to vary with environmental conditions (namely, temperature and crowding) in experimental populations of the milkweed bug (*Oncopeltus fasciatus*) (Dingle, 1968). Frank *et al.* (1957) showed changes in survivorship related to differences in density of laboratory populations of *Daphnia pulex*, and Root and Olson (1969) showed variations in survivorship of cabbage aphid (*Brevicoryne brassicae*) populations growing on different host plants. These temporal variations in survivorship result in field populations having nonstable age distributions, the effects of which on numbers (Fig. 2.3) must be accounted for when one is developing hypotheses regarding the causal factors determining the size and fluctuations of a population's numbers.

It is now time to review the theory of population dynamics presented in this book by answering the questions posed in Chapter 1.

1. Why does the sample population grow to a particular size instead of some larger or smaller size?

A population's maximum size is limited by the availability of resources (food, space, or time) or by predators or by disease in the particular spatial unit being studied. Food-limited populations are limited by the amount of food that is available to them (Fig. 3.5), which is a function not only of the total amount present but of the animals' abilities to find, capture, and metabolize the food. It may be useful to distinguish the absolute amount from the relative amount of food in a particular case (Andrewartha and Browning, 1961), but what is important are the probabilities of surviving and reproducing of individuals given a particular supply of food. At higher densities, increasing intraspecific competition for food results in higher mortality and lower fecundity, until the growth rate finally becomes zero or negative.

The maximum population of space-limited populations is set by the minimal size of the space occupied by reproductive individuals or pairs and the amount of suitable space (Fig. 3.1). All populations are potentially food-limited or space-limited. Sometimes, however, they do not reach the maximum size allowed by the available resources. These populations may not have had sufficient time in which to grow (Fig. 3.13), or they may be predator- or disease-limited (Figs. 3.8, 5.3, and 5.13).

In none of these cases does it seem necessary to invoke density-dependent factors as regulators of the size of populations.

2. Why does the population's size fluctuate about some mean value instead of remaining constant or fluctuating even more?

Except for space-limited populations (Fig. 3.1), a steady-state population size is a virtual impossibility. Food-limited populations consume their food supply, reducing their size and availability, and they are dependent on the rate at which food is replenished. Unless the replenishment rate equals the rate of consumption (which seems unlikely), the population's size will fluctuate as the consumer and consumed populations alternately depress each other's growth rate (Fig. 3.8).

Some populations will fluctuate because the length of the breeding season changes with the vagaries of the physical environment (Fig. 3.13). Space-limited populations fluctuate in size rather than remain constant because factors other than the minimal territory size are active in determining population numbers. For example, the minimal territory size changes in response to environmental factors (Tinbergen, 1957; Brown, 1969a; Watson and Moss, 1970; Klomp, 1972). Periods of fluctuating population sizes can be caused by a one-time change in recruitment (Fig. 2.3). A series of changes in recruitment because of alternating good and poor breeding seasons will make predicting the future size of a population risky unless the predictor has much more information about the population than ecologists usually have.

The magnitude of the fluctuations of food-limited and predator-limited populations depends on the inherent characteristics of the individuals of both consumer and consumed populations, such as the hunting ability of the predator, the crypticity of the prey, and the dispersal capabilities of both predator and prey, and on the number of populations interacting within the community. A rapidly growing, efficiently hunting predator is likely to have a greater impact on its food supply and induce greater fluctuations in its own population than a more slowly growing, less efficient predator. A community with more links between predators and prey should have smaller fluctuations than one with fewer predators and prey [MacArthur (1955) and Leigh (1965); but see Hairston et al. (1968) and May (1973)].

3. Why does a population at one place differ in size and in the amplitude

of its fluctuations from another population of the same species in another area?

Populations differ in size from place to place because the quality of the environment (both physical and biotic) varies and affects the values of the birth and replacement rates (Fig. 3.19), because of differences in the lengths of time of the reproductive and nonreproductive seasons (Figs. 3.13 and 3.14), and because of the degree of predation (Fig. 5.3). The presence or absence of competitors can affect a population's numbers (Figs. 6.2–6.4), although these effects will be transitory unless other factors ameliorate the intensity of competition, as discussed in Chapter 6. Geographic variation in the amplitude of the fluctuations can be induced by variation in the magnitude of environmental effects on recruitment (Fig. 2.3).

4. Are the factors that determine the size of a population and the amplitude of its fluctuations in one area the same as those that determine the size and amplitude of fluctuations of a population in a larger area?

This cannot be answered theoretically (at least not by the theory developed in this book). Intuitively, there seems no reason to assume that the dynamics of different populations are either necessarily the same or necessarily different. Thus, each population must be examined directly.

Empirically, however, we have seen that the size of the study area does affect interpretation of the dynamics of the populations within it. The population within a given area may often consist of subunits, and the dynamics of the subunits can be different from the dynamics of the whole, as already discussed for predator–prey interactions of *Cactoblastis* and *Opuntia* and of *Typhlodromus* and *Eotetranychus* (Chapter 1), for territorial behavior of birds (Chapter 3), and for competitive interactions (Chapter 6).

5. Are the factors that determine the commonness or rarity (abundance) of a species within the sample areas the same as those that determine the extent of the species' geographic range (distribution)?

The size of any population depends on (a) the maximum birth rate and minimum replacement rate that a population can achieve in a particular environment and (b) the length of time that a positive growth rate can be maintained. The latter is determined by the availability of space (Fig. 3.1), of food (Fig. 3.5), or of time itself (Fig. 3.13). The former are a consequence of the morphological, physiological, and behavioral attributes of the members of the population, which have evolved by natural selection (Chapter 4), and of the prevailing environmental conditions, including predators or pathogens (Chapter 5) and competitors (Chapter 6). Of two populations, the one with the larger growth rate and the longer period of positive growth will attain a larger population size (Fig. 3.13). Conditions vary geographically, affecting the population's maximum potential growth rate and the length of time the growth rate is positive, and therefore population densities will vary geographically (Fig. 3.19). The species will not occur wherever the replacement rate exceeds the birth rate. It seems, then, that the physical and biotic components of the environment determine both the abundance and distribution of particular species. There is no difference in mechanism, although the particular features of the environment affecting a population's survivorship and fecundity schedules may vary from place to place and from time to time.

Zoogeographers map past and present distributions of species and higher taxa, discover patterns in these distributions, and hypothesize historical events accounting for changes in the ranges, abundance, diversity, and extinction of groups (Darlington, 1957). Important parameters affecting the distributions of both animals and plants are temperature and rainfall and the occurrence of competitors. Physiologists obtain data on the responses of animals to a variety of environmental factors, including temperature, humidity, pH, and salinity. Natural historians and population ecologists collect data on survivorship and fecundity of animals under various conditions. Yet how the components of the environment—weather, food, other animals, and shelter (Andrewartha and Birch, 1954)—affect survivorship and fecundity are virtually unknown, for at least two reasons. First, acquiring such data from natural populations is extraordinarily difficult. Environmental conditions are not constant, and therefore individual probabilities of surviving and reproducing are changing, resulting in nonstable, non-steady-state populations. Second, the theory of population dynamics prevailing during the past quarter-century emphasized density-dependent factors and encouraged analysis of series of population counts rather than stimulated research on those factors affecting survivorship and fecundity. The interest in survivorship and fecundity data is in describing life history patterns and in providing evolutionary explanations for them.

With the wealth of information now available, it is time for synthesis, and this book represents one beginning. Because the problems discussed in this book are usually approached separately, and usually from different points of view, important data are lacking. Some examples include: survivorship schedules from birth are rare; when clutch size is well known, as in birds, the number of clutches per season is poorly known; the upper critical densities or the minimal acceptable territory sizes are unknown and may never be known. Yet, despite their limitations, the available data seem consistent with the predictions of the theories. From the point of view of the models presented here, however, much research remains to be done.

BIBLIOGRAPHY

Allee, W. C., Emerson, A. E., Park, O., Park, T., and Schmidt, K. P. (1949). "Principles of Animal Ecology." Saunders, Philadelphia, Pennsylvania.

Amadon, D. (1964). The evolution of low reproductive rates in birds. *Evolution* **18**, 105–110.

Anderson, P. K. (1960). Ecology and evolution in island populations of salamanders in the San Francisco Bay region. *Ecol. Monogr.* **30**, 359–385.

Andrewartha, H. G. (1958). The use of conceptual models in population ecology. *Cold Spring Harbor Symp. Quant. Biol.* **22**, 219–236.

Andrewartha, H. G. (1959). Density-dependent factors in ecology. *Nature (London)* **183**, 200.

Andrewartha, H. G. (1963). Density-dependence in the Australian thrips. *Ecology* **44**, 218–220.

Andrewartha, H. G., and Birch, L. C. (1954). "The Distribution and Abundance of Animals." Univ. of Chicago Press, Chicago, Illinois.

Andrewartha, H. G., and Birch, L. C. (1960). Some recent contributions to the study of the distribution and abundance of insects. *Annu. Rev. Entomol.* **5**, 219–242.

Andrewartha, H. G., and Browning, T. O. (1961). An analysis of the idea of "resources" in animal ecology. *J. Theor. Biol.* **1**, 83–97.

Asdell, S. A. (1964). "Patterns of Mammalian Reproduction." Cornell Univ. Press, Ithaca, New York.

Ashmole, N. P. (1963). The regulation of numbers of tropical oceanic birds. *Ibis* **103b**, 458–473.

Ashmole, N. P. (1971). Sea bird ecology and the marine environment. *In* "Avian Biology" (D. S. Farner and J. R. King, eds.), Vol. 1, pp. 223–286. Academic Press, New York.

Ayala, F. J. (1969). Experimental invalidation of the principle of competitive exclusion. *Nature (London)* **224**, 1076–1079.

Bakker, K. (1964). Backgrounds of controversies about population theories and their terminologies. *Z. Angew. Entomol.* **53**, 187–208.

Bakker, K. (1970). Some general remarks on the concepts "population" and "regulation." *In*

"Dynamics of Populations" (P. J. den Boer and G. R. Gradwell, eds.), pp. 565–567. Cent. Agric. Publ. Docum., Wageningen.

Bartlett, M. S. (1973). Equations and models of population change. *In* "The Mathematical Theory of the Dynamics of Biological Populations" (M. S. Bartlett and R. W. Hiorns, eds.), pp. 5–21. Academic Press, New York.

Beauchamp, R. S. A., and Ullyott, P. (1932). Competitive relationships between certain species of fresh-water triclads. *J. Ecol.* **20**, 200–208.

Beddington, J. R., and May, R. M. (1977). Harvesting natural populations in a randomly fluctuating environment. *Science* **197**, 463–465.

Berger, M. E. (1972). Population structure of olive baboons [*Papio anubis* (J. P. Fischer)] in the Laikipia District of Kenya. *East Afr. Wildl. J.* **10**, 159–164.

Beverton, R. J. H., and Holt, S. J. (1957). "On the Dynamics of Exploited Fish Populations." HM Stationery Off., London.

Birch, L. C. (1948). The intrinsic rate of natural increase of an insect population. *J. Anim. Ecol.* **17**, 15–26.

Birch, L. C. (1957). The meanings of competition. *Am. Nat.* **91**, 5–18.

Birch, L. C. (1958). The role of weather in determining the distribution and abundance of animals. *Cold Spring Harbor Symp. Quant. Biol.* **22**, 203–218.

Birch, L. C., and Ehrlich, P. R. (1967). Evolutionary history and population biology. *Nature (London)* **214**, 349–352.

Bonner, J. T. (1965). "Size and Cycle: An Essay on the Structure of Biology." Princeton Univ. Press, Princeton, New Jersey.

Boulding, K. E. (1955). An application of population analysis to the automobile population of the United States. *Kyklos* **8**, 109–124.

Brockelman, W. Y. (1975). Competition, the fitness of offspring, and optimal clutch size. *Am. Nat.* **109**, 677–699.

Bronowski, J. (1965). "Science and Human Values," Rev. Ed. Harper, New York.

Brown, J. L. (1964). The evolution of diversity in avian territorial systems. *Wilson Bull.* **76**, 160–169.

Brown, J. L. (1969a). Territorial behavior and population regulation in birds. A review and re-evaluation. *Wilson Bull.* **81**, 293–329.

Brown, J. L. (1969b). The buffer effect and productivity in tit populations. *Am. Nat.* **103**, 347–354.

Brown, W. L., Jr., and Wilson, E. O. (1956). Character displacement. *Syst. Zool.* **5**, 49–64.

Bryant, D. M. (1975). Breeding biology of House Martins *Delichon urbica* in relation to aerial insect abundance. *Ibis* **117**, 180–216.

Bryant, E. H. (1971). Life history consequences of natural selection: Cole's result. *Am. Nat.* **105**, 75–76.

Buckner, C. H., and Turnock, W. J. (1965). Avian predation on the Larch Sawfly, *Pristiphora erichsonii* (Htg.), (Hymenoptera: Tenthredinidae). *Ecology* **46**, 223–236.

Cagle, F. R. (1950). The life history of the slider turtle, *Pseudemys scripta troostii* (Holbrook). *Ecol. Monogr.* **20**, 31–54.

Cagle, F. R. (1954). Observations on the life cycles of painted turtles (genus *Chrysemys*). *Am. Midl. Nat.* **52**, 225–235.

Carrick, R. (1963). Ecological significance of territory in the Australian Magpie, *Gymnorhina tibicen. Proc. Int. Ornithol. Congr., 13th, 1962* pp. 740–753.

Caughley, G. (1966). Mortality patterns in mammals. *Ecology* **47**, 906–918.

Caughley, G. (1967). Parameters for seasonally breeding populations. *Ecology* **48**, 834–839.

Chapman, R. N. (1928). The quantitative analysis of environmental factors. *Ecology* **9**, 111–122.

Charnov, E. L., and Krebs, J. R. (1974). On clutch-size and fitness. *Ibis* **116**, 217–219.

Charnov, E. L., and Schaffer, W. M. (1973). Life-history consequences of natural selection: Cole's result revisited. *Am. Nat.* **107**, 791–793.

Chitty, D. (1960). Population processes in the vole and their relevance to general theory. *Can. J. Zool.* **38**, 99–113.

Chitty, D. (1967). The natural selection of self-regulatory behavior in animal populations. *Proc. Ecol. Soc. Aust.* **2**, 51–78.

Clark, C. W. (1973). The economics of overexploitation. *Science* **181**, 630–634.

Clark, C. W. (1976). "Mathematical Bioeconomics: The Optimal Management of Renewable Resources." Wiley, New York.

Clark, L. R., Geier, P. W., Hughes, R. D., and Morris, R. F. (1967). "The Ecology of Insect Populations in Theory and Practice." Methuen, London.

Claude, C. (1970). Biometrie und Fortpflanzungsbiologie der Rötelmaus *Clethrionomys glareolus* (Schreber, 1780) auf verschiedenen Höhenstufen der Schweiz. *Rev. Suisse Zool.* **77**, 435–480.

Coale, A. J. (1958). How the age distribution of a human population is determined. *Cold Spring Harbor Symp. Quant. Biol.* **22**, 83–89.

Cody, M. L. (1966). A general theory of clutch size. *Evolution* **20**, 174–184.

Cody, M. L. (1969). The evolution of reproductive rates (review of D. Lack's "Ecological Adaptations for Breeding in Birds"). *Science* **163**, 1185–1187.

Cody, M. L. (1971). Ecological aspects of reproduction. *In* "Avian Biology" (D. S. Farner and J. R. King, eds.), Vol. 1, pp. 461–512. Academic Press, New York.

Cole, L. C. (1954). The population consequences of life history phenomena. *Q. Rev. Biol.* **29**, 103–137.

Cole, L. C. (1960). Competitive exclusion. *Science* **132**, 348–349.

Connell, J. H. (1961). The influence of interspecific competition and other factors on the distribution of the barnacle *Chthamalus stellatus*. *Ecology* **42**, 710–723.

Corkhill, P. (1973). Food and feeding ecology of Puffins. *Bird Study* **20**, 207–220.

Coulson, J. C. (1956). Mortality and egg production of the Meadow Pipit with special reference to altitude. *Bird Study* **3**, 119–132.

Crombie, A. C. (1947). Interspecific competition. *J. Anim. Ecol.* **16**, 44–73.

Crossner, K. A. (1977). Natural selection and clutch size in the European Starling. *Ecology* **58**, 885–892.

Curio, E. (1958). Geburtsortstreue und Lebenserwartung junger Trauerschnäpper (*Muscicapa h. hypoleuca* Pallas). *Vogelwart* **79**, 135–148.

Curtis, J. T., and McIntosh, R. P. (1950). The interrelations of certain analytic and synthetic phytosociological characters. *Ecology* **31**, 434–455.

Darlington, P. J., Jr. (1957). "Zoogeography: The Geographical Distribution of Animals." Wiley, New York.

Darwin, C. (1872). "The Origin of Species" (Reprint of 6th Ed.). Random House, New York.

Davidson, J., and Andrewartha, H. G. (1948a). Annual trends in a natural population of *Thrips imaginis* (Thysanoptera). *J. Anim. Ecol.* **17**, 193–199.

Davidson, J., and Andrewartha, H. G. (1948b). The influence of rainfall, evaporation and atmospheric temperature on fluctuations in the size of a natural population of *Thrips imaginis* (Thysanoptera). *J. Anim. Ecol.* **17**, 200–222.

Davies, N. B. (1977). Prey selection and search strategy of the Spotted Flycatcher (*Muscicapa striata*): A field study on optimal foraging. *Anim. Behav.* **25**, 1016–1033.

DeBach, P. (1966). The competitive displacement and coexistence principles. *Annu. Rev. Entomol.* **11**, 183–212.

Deevey, E. S., Jr. (1947). Life tables for natural populations of animals. *Q. Rev. Biol.* **22**, 283–314.

Delius, J. D. (1965). A population study of Skylarks (*Alauda arvensis*). *Ibis* **107**, 466–492.

Dementiev, G. P., and Stépanyan, L. S. (1965). Quelques particularites de la reproduction des oiseaux dans les hautes zones du Thian-Chan. *Oiseau* **35**, Spec. No., 58–64.

DeWitt, R. M. (1954). The intrinsic rate of natural increase in a pond snail (*Physa gyrina* Say). *Am. Nat.* **88**, 353–359.

Dingle, H. (1968). Life history and population consequences of density, photo-period, and temperature in a migrant insect, the milkweed bug *Oncopeltus. Am. Nat.* **102**, 149–163.

Dodd, A. P. (1959). The biological control of the prickly pear in Australia. *In* "Biogeography and Ecology in Australia" (A. Keast, R. L. Crocker, and C. S. Christian, eds.), Monographiae Biologicae, Vol. 8, pp. 565–577. Junk, The Hague.

Dorwood, D. F. (1962). Comparative biology of the White Booby and Brown Booby *Sula* spp. at Ascension. *Ibis* **103b**, 174–220.

Downes, J. A. (1964). Arctic insects and their environment. *Can. Entomol.* **96**, 279–307.

Dublin, L. I., and Lotka, A. J. (1925). On the true rate of natural increase. *J. Am. Statist. Assoc.* **20**, 305–339.

Dunmire, W. W. (1960). An altitudinal survey of reproduction in *Peromyscus maniculatus. Ecology* **41**, 174–182.

Eberhardt, L. L. (1977). Optimal policies for conservation of large mammals, with special reference to marine ecosystems. *Environ. Conserv.* **4**, 205–212.

Eberhardt, L. L., and Siniff, D. B. (1977). Population dynamics and marine mammal management policies. *J. Fish. Res. Board Can.* **34**, 183–190.

Ehrlich, P. R., and Birch, L. C. (1967). The "balance of nature" and "population control." *Am. Nat.* **101**, 97–107.

Ehrlich, P. R., Breedlove, D. E., Brussard, P. F., and Sharp, M. A. (1972). Weather and the "regulation" of subalpine populations. *Ecology* **53**, 243–247.

Elliott, P. F. (1975). Longevity and the evolution of polygamy. *Am. Nat.* **109**, 281–287.

Elton, C. (1942). "Voles, Mice and Lemmings. Problems in Population Dynamics." Oxford Univ. Press, London.

Emlen, J. M. (1966). The role of time and energy in food preference. *Am. Nat.* **100**, 611–617.

Errington, P. L. (1945). Some contributions of a fifteen-year local study of the Northern Bobwhite to a knowledge of population phenomena. *Ecol. Monogr.* **15**, 1–34.

Errington, P. L. (1946). Predation and vertebrate populations. *Q. Rev. Biol.* **21**, 144–177, 221–245.

Evans, F. C. (1952). The influence of size of quadrat on the distributional patterns of plant populations. *Contrib. Lab. Vertebr. Biol. Univ. Mich.* **54**, 1–15.

Evans, F. C., and Smith, F. E. (1952). The intrinsic rate of natural increase for the human louse, *Pediculus humanus* L. *Am. Nat.* **86**, 299–310.

Farner, D. S. (1955). Birdbanding in the study of population dynamics. *In* "Recent Studies in Avian Biology" (A. Wolfson, ed.), pp. 397–449. Univ. of Illinois Press, Urbana.

Feller, W. (1940). On the logistic law of growth and its empirical verifications in biology. *Acta Biotheor.* **5**, 51–66.

Ferry, C., and Deschaintre, A. (1974). Le chant, signal interspécifique chez *Hippolais icterina* et *polyglotta. Alauda* **42**, 289–311.

Fisher, R. A. (1930). "The Genetical Theory of Natural Selection." Oxford Univ. Press, London.

Fotheringham, N. (1971). Life history patterns of the littoral gastropods *Shaskyus festivus* (Hinds) and *Ocenebra poulsoni* Carpenter (Prosobranchia: Muricidae). *Ecology* **52**, 742–757.

Frank, P. W., Boll, C. D., and Kelly, R. W. (1957). Vital statistics of laboratory cultures of *Daphnia pulex* DeGeer as related to density. *Physiol. Zool.* **30**, 287–305.

Fretwell, S. D. (1969). The adjustment of birth rate to mortality in birds. *Ibis* **111**, 624–627.

Fretwell, S. D., and Calver, J. S. (1970). On territorial behavior and other factors influencing habitat distribution in birds. II. Sex ratio variation in the Dickcissel (*Spiza americana* Gmel.). *Acta Biotheor.* **19**, 37–44.

Fry, C. H. (1977). The evolutionary significance of co-operative breeding in birds. *In* "Evolutionary Ecology" (B. Stonehouse and C. Perrins, eds.), pp. 127–135. Univ. Park Press, Baltimore, Maryland.

Fuggles-Couchman, N. R. (1943). A contribution to the breeding ecology of two species of *Euplectes* (Bishop-birds) in Tanganyika Territory. *Ibis* **85**, 311–326.

Gause, G. F. (1934). "The Struggle for Existence." Williams & Wilkins, Baltimore, Maryland.

Gause, G. F., and Witt, A. A. (1935). Behavior of mixed populations and the problem of natural selection. *Am. Nat.* **69**, 596–609.

Giesel, J. T. (1976). Reproductive strategies as adaptations to life in temporally heterogeneous environments. *Annu. Rev. Ecol. Syst.* **7**, 57–79.

Gilbert, O., Reynoldson, T. B., and Hobart, J. (1952). Gause's hypothesis: An examination. *J. Anim. Ecol.* **21**, 310–312.

Glander, K. E. (1977). Poison in a monkey's Garden of Eden. *Nat. Hist.* **86**(3), 35–41.

Gleason, H. A. (1922). On the relation between species and area. *Ecology* **3**, 158–162.

Goss-Custard, J. D. (1977a). The energetics of prey selection by Redshank, *Tringa totanus* (L.), in relation to prey density. *J. Anim. Ecol.* **46**, 1–19.

Goss-Custard, J. D. (1977b). Optimal foraging and the size selection of worms by Redshank, *Tringa totanus*, in the field. *Anim. Behav.* **25**, 10–29.

Graham, M. (1935). Modern theory of exploiting a fishery, and application to North Sea trawling. *J. Cons., Cons. Perm. Int. Explor. Mer* **10**, 264–274.

Grant, P. R. (1972). Convergent and divergent character displacement. *Biol. J. Linn. Soc.* **4**, 39–68.

Gray, J. (1929). The kinetics of growth. *Br. J. Exp. Biol.* **6**, 248–274.

Griffiths, J. T., Jr., and Tauber, O. E. (1942). Fecundity, longevity, and parthenogenesis of the American roach, *Periplaneta americana* L. *Physiol. Zool.* **15**, 196–209.

Gulland, J. A. (1962). The application of mathematical models to fish populations. *In* "The Exploitation of Natural Animal Populations" (E. D. Le Cren and M. W. Holdgate, eds.), British Ecological Society Symposium, No. 2, pp. 204–217. Wiley, New York.

Hairston, N. G., Allan, J. D., Colwell, R. K., Futuyma, D. J., Howell, J., Lubin, M. D., Mathias, J., and Vandermeer, J. H. (1968). The relationship between species diversity and stability: An experimental approach with protozoa and bacteria. *Ecology* **49**, 1091–1101.

Hairston, N. G., Tinkle, D. W., and Wilbur, H. M. (1970). Natural selection and the parameters of population growth. *J. Wildl. Manage.* **34**, 681–690.

Hamilton, W. D. (1966). The moulding of senescence by natural selection. *J. Theor. Biol.* **12**, 12–45.

Hardin, G. (1960). The competitive exclusion principle. *Science* **131**, 1292–1297.

Hardin, G. (1968). The tragedy of the commons. *Science* **162**, 1243–1248.

Harris, M. P. (1966). Breeding biology of the Manx Shearwater *Puffinus puffinus. Ibis* **108**, 17–33.

Harris, M. P. (1970a). Territory limiting the size of the breeding population of the Oyster-catcher (*Haematopus ostralegus*)—A removal experiment. *J. Anim. Ecol.* **39**, 707–713.

Harris, M. P. (1970b). Breeding ecology of the Swallow-tailed Gull, *Creagrus furcatus. Auk* **87**, 215–243.

Harris, M. P., and Plumb, W. J. (1965). Experiments on the ability of Herring Gulls *Larus argentatus* and Lesser Black-backed Gulls *L. fuscus* to raise larger than normal broods. *Ibis* **107**, 256–257.

Haymes, G. T., and Morris, R. D. (1977). Brood size manipulations in herring gulls. *Can. J. Zool.* **55**, 1762–1766.

Healey, M. C. (1967). Aggression and self-regulation of population size in deermice. *Ecology* **48**, 377–392.

Hensley, M. M., and Cope, J. B. (1951). Further data on removal and repopulation of the breeding birds in a spruce–fir forest community. *Auk* **68**, 483–493.

Heron, A. C. (1972). Population ecology of a colonizing species: The pelagic tunicate *Thalia democratica*. II. Population growth rate. *Oecologia* **10**, 294–312.

Hickey, J. J. (1952). Survival studies of banded birds. *U.S. Fish and Wildl. Serv., Spec. Sci. Rep.: Wildl.* No. 15.

Hinde, R. A. (1956). The biological significance of the territories of birds. *Ibis* **98**, 340–369.

Hogben, L. (1931). Some biological aspects of the population problem. *Biol. Rev. Cambridge Philos. Soc.* **6**, 163–180.

Holgate, P. (1967). Population survival and life history phenomena. *J. Theor. Biol.* **14**, 1–10.

Holling, C. S. (1959). The components of predation as revealed by a study of small-mammal predation of the European pine sawfly. *Can. Entomol.* **91**, 293–320.

Holmes, R. T. (1966). Breeding ecology and annual cycle adaptations of the Red-backed Sandpiper (*Calidris alpina*) in northern Alaska. *Condor* **68**, 3–46.

Holyoak, D. (1967). Breeding biology of the Corvidae. *Bird Study* **14**, 153–168.

Howard, L. O., and Fiske, W. F. (1912). The importation into the United States of the parasites of the gypsy moth and the brown-tail moth. *Bull. U.S. Bur. Entomol.* No. 91.

Huffaker, C. B. (1958a). The concept of balance in nature. *Proc. Int. Congr. Entomol., 10th* **2**, 625–636.

Huffaker, C. B. (1958b). Experimental studies on predation: Dispersion factors and predator-prey oscillations. *Hilgardia* **27**, 343–383.

Huffaker, C. B., and Messenger, P. S. (1964a). The concept and significance of natural control. *In* "Biological Control of Insect Pests and Weeds" (P. DeBach, ed.), pp. 74–117. Reinhold, New York.

Huffaker, C. B., and Messenger, P. S. (1964b). Population ecology—Historical development. *In* "Biological Control of Insect Pests and Weeds" (P. DeBach, ed.), pp. 45–73. Reinhold, New York.

Hussell, D. J. T. (1972). Factors affecting clutch size in arctic passerines. *Ecol. Monogr.* **42**, 317–364.

Hutchinson, G. E. (1948). Circular causal systems in ecology. *Ann. N.Y. Acad. Sci.* **50**, 221–246.

Hutchinson, G. E. (1958). Concluding remarks. *Cold Spring Harbor Symp. Quant. Biol.* **22**, 415–427.

Hutchinson, G. E. (1959). Homage to Santa Rosalia *or* why are there so many kinds of animals? *Am. Nat.* **93**, 145–159.

Huxley, J. S. (1934). A natural experiment on the territorial instinct. *Br. Birds* **27**, 270–277.

Idyll, C. P. (1973). The anchovy crisis. *Sci. Am.* **228**, 22–29.

Jarvis, M. J. F. (1974). The ecological significance of clutch size in the South African Gannett *Sula capensis* (Lichtenstein). *J. Anim. Ecol.* **43**, 1–17.

Jehl, J. R., Jr. (1970). Fish, fowl, and fertilizer on the Peruvian coast: Problems for survival ecology. *Environ. Southwest* Aug./Sept., pp. 3–5.

Jehl, J. R., Jr., and Hussell, D. J. T. (1966). Effects of weather on reproductive success of birds at Churchill, Manitoba. *Arctic* **19**, 185–191.

Johnson, N. K. (1963). Biosystematics of sibling species of flycatchers in the *Empidonax hammondii-oberholseri-wrightii* complex. *Univ. Calif. Berkeley Publ. Zool.* **66**(2), 79–238.

Johnson, N. K. (1966). Bill size and the question of competition in allopatric and sympatric populations of Dusky and Gray flycatchers. *Syst. Zool.* **15**, 70–87.

Johnston, R. F. (1954). Variation in breeding season and clutch size in Song Sparrows of the Pacific coast. *Condor* **56**, 268–273.

Kavanagh, A. J., and Richards, O. W. (1934). The autocatalytic growth-curve. *Am. Nat.* **68**, 54–59.

Keith, L. B. (1963). "Wildlife's Ten-Year Cycle." Univ. of Wisconsin Press, Madison.

Kendeigh, S. C. (1941). Territorial and mating behavior of the House Wren. *Ill. Biol. Monogr.* **18**, 1–120.

King, C. E. (1964). Relative abundance of species and MacArthur's model. *Ecology* **45**, 716–727.

Klomp, H. (1962). The influence of climate and weather on the mean density level, the fluctuations and the regulation of animal populations. *Arch. Neerl. Zool.* **15**, 68–109.

Klomp, H. (1970). The determination of clutch-size in birds. A review. *Ardea* **58**, 1–124.

Klomp, H. (1972). Regulation of the size of bird populations by means of territorial behaviour. *Neth. J. Zool.* **22**, 456–488.

Kluyver, H. N., and Tinbergen, L. (1953). Territory and the regulation of density in titmice. *Arch. Neerl. Zool.* **10**, 265–287.

Kramer, G. (1946). Veränderungen von Nachkommenziffer und Nachkommengrösse sowie der Altersverteilung von Inseleidechsen. *Z. Naturforsch.* **1**, 700–710.

Krebs, J. R. (1970). Regulation of numbers in the Great Tit (Aves: Passeriformes). *J. Zool.* **162**, 317–333.

Krebs, J. R. (1971). Territory and breeding density in the Great Tit, *Parus major* L. *Ecology* **52**, 2–22.

Krebs, J. R., and Cowie, R. J. (1976). Foraging strategies in birds. *Ardea* **64**, 98–116.

Krebs, J. R., Erichsen, J. T., Webber, M. I., and Charnov, E. L. (1977). Optimal prey selection in the Great Tit *(Parus major)*. *Anim. Behav.* **25**, 30–38.

Kuenen, D. J. (1958). Some sources of misunderstanding in the theories of regulation of animal numbers. *Arch. Neerl. Zool.* **13**, Suppl. 1, 335–341.

Lack, D. (1944). Ecological aspects of species-formation in passerine birds. *Ibis* **86**, 260–286.

Lack, D. (1945). The ecology of closely related species with special reference to Cormorant *(Phalacrocorax carbo)* and Shag *(P. aristotelis)*. *J. Anim. Ecol.* **14**, 12–16.

Lack, D. (1946). Competition for food by birds of prey. *J. Anim. Ecol.* **15**, 123–129.

Lack, D. (1947a). The significance of clutch-size. Parts I and II. *Ibis* **89**, 302–352.

Lack, D. (1947b). "Darwin's Finches." Cambridge Univ. Press, London and New York.

Lack, D. (1948a). The significance of clutch-size. Part III. *Ibis* **90**, 25–45.

Lack, D. (1948b). The significance of litter-size. *J. Anim. Ecol.* **17**, 45–50.

Lack, D. (1948c). Natural selection and family size in the Starling. *Evolution* **2**, 95–110.

Lack, D. (1949). Family size in certain thrushes (Turdidae). *Evolution* **3**, 57–66.

Lack, D. (1954). "The Natural Regulation of Animal Numbers." Oxford Univ. Press, London and New York.

Lack, D. (1966). "Population Studies of Birds." Oxford Univ. Press, London and New York.

Lack, D. (1968). "Ecological Adaptations for Breeding in Birds." Methuen, London.

Lack, D. (1971). "Ecological Isolation in Birds." Harvard Univ. Press, Cambridge, Massachusetts.

Lack, D., and Lack, E. (1951). The breeding biology of the swift *Apus apus*. *Ibis* **93**, 501–546.

Lack, D., Gibb, J., and Owen, D. F. (1957). Survival in relation to brood-size in tits. *Proc. Zool. Soc. London* **128**, 313–326.

Lanyon, W. E. (1956). Territory in the meadowlarks, genus *Sturnella*. *Ibis* **98**, 485–489.

Lanyon, W. E. (1957). The comparative biology of the meadowlarks (*Sturnella*) in Wisconsin. *Nuttall Ornithol. Club Publ.* No. 1.

Larkin, P. A. (1977). An epitaph for the concept of maximum sustainable yield. *Trans. Am. Fish. Soc.* **106**, 1–11.

Lawler, G. H. (1965). Fluctuations in the success of year-classes of whitefish populations with special reference to Lake Erie. *J. Fish. Res. Board Can.* **22**, 1197–1227.

Leigh, E. G., Jr. (1965). On the relation between the productivity, biomass, diversity, and stability of a community. *Proc. Natl. Acad. Sci. U.S.A.* **53**, 777–783.

Leslie, P. H. (1945). On the use of matrices in certain population mathematics. *Biometrika* **33**, 183–212.

Leslie, P. H. (1948). Some further notes on the use of matrices in population mathematics. *Biometrika* **35**, 213–245.

Leslie, P. H. (1966). The intrinsic rate of increase and the overlap of successive generations in a population of guillemots (*Uria aalge* Pont.). *J. Anim. Ecol.* **35**, 291–301.

Leslie, P. H., and Park, T. (1949). The intrinsic rate of natural increase of *Tribolium castaneum* Herbst. *Ecology* **30**, 469–477.

Leslie, P. H., and Ranson, R. M. (1940). The mortality, fertility and rate of natural increase of the vole (*Microtus agrestis*) as observed in the laboratory. *J. Anim. Ecol.* **9**, 27–52.

Lidicker, W. Z., Jr. (1962). Emigration as a possible mechanism permitting the regulation of population density below carrying capacity. *Am. Nat.* **96**, 29–33.

Lill, A. (1974). The evolution of clutch size and male "chauvinism" in the White-bearded Manakin. *Living Bird* **13**, 211–231.

Lloyd, C. S. (1977). The ability of the Razorbill *Alca torda* to raise an additional chick to fledging. *Ornis Scand.* **8**, 155–159.

Lord, R. D., Jr. (1960). Litter size and latitude in North American mammals. *Am. Midl. Nat.* **64**, 488–499.

Lotka, A. J. (1907a). Relation between birth rates and death rates. *Science* **26**, 21–22.

Lotka, A. J. (1907b). Studies on the mode of growth of material aggregates. *Am. J. Sci.* **24**, 199–216.

Lotka, A. J. (1913a). A natural population norm. I. *J. Wash. Acad. Sci.* **3**, 241–248.

Lotka, A. J. (1913b). A natural population norm. II. *J. Wash. Acad. Sci.* **3**, 289–293.

Lotka, A. J. (1922). The stability of the normal age distribution. *Proc. Natl. Acad. Sci. U.S.A.* **8**, 339–345.

Lotka, A. J. (1925). "Elements of Physical Biology." Williams & Wilkins, Baltimore, Maryland.

Lotka, A. J. (1927). The size of American families in the eighteenth century, and the significance of the empirical constants in the Pearl–Reed law of population growth. *J. Am. Statist. Assoc.* **22**, 154–170.

Lotka, A. J. (1932). The growth of mixed populations: Two species competing for a common food supply. *J. Wash. Acad. Sci.* **22**, 461–469.

Lotka, A. J. (1943). The place of the intrinsic rate of natural increase in population analysis. *Proc. Am. Sci. Congr., 8th, 1940* **8**, 297–313.

MacArthur, R. (1955). Fluctuations of animal populations, and a measure of community stability. *Ecology* **36**, 533–536.

MacArthur, R. H., and Pianka, E. R. (1966). On optimal use of a patchy environment. *Am. Nat.* **100**, 603–609.

MacArthur, R. H., and Wilson, E. O. (1967). "The Theory of Island Biogeography." Princeton Univ. Press, Princeton, New Jersey.

Maher, W. J. (1967). Predation by weasels on a winter population of lemmings, Banks Island, Northwest Territories. *Can. Field Nat.* **81**, 248–250.

Maher, W. J. (1970). The Pomarine Jaeger as a brown lemming predator in northern Alaska. *Wilson Bull.* **82**, 130–157.

Maher, W. J. (1974). Ecology of Pomarine, Parasitic, and Long-tailed jaegers in northern Alaska. *Pac. Coast Avifauna* No. 37.

Maiorana, V. C. (1976). Reproductive value, prudent predators, and group selection. *Am. Nat.* **110**, 486–489.

Margalef, R. (1968). "Perspectives in Ecological Theory." Univ. of Chicago Press, Chicago, Illinois.

Markus, H. C. (1934). Life history of the Black-headed Minnow (*Pimephales promelas*). *Copeia* pp. 116–122.

May, R. M. (1973). "Stability and Complexity in Model Ecosystems." Princeton Univ. Press, Princeton, New Jersey.

May, R. M. (1976a). Models for single populations. *In* "Theoretical Ecology: Principles and Applications" (R. M. May, ed.), pp. 4–25. Saunders, Philadelphia, Pennsylvania.

May, R. M., ed. (1976b). "Theoretical Ecology: Principles and Applications." Saunders, Philadelphia, Pennsylvania.

Maynard Smith, J. (1959). The rate of ageing in *Drosophila subobscura*. *In* "The Lifespan of Animals" (G. E. W. Wolstenholme and M. O'Connor, eds.), CIBA Foundation Colloquia on Ageing, Vol. 5, pp. 269–281. Little, Brown, Boston, Massachusetts.

Maynard Smith, J., and Slatkin, M. (1973). The stability of predator–prey systems. *Ecology* **54**, 384–391.

Mayr, E. (1963). "Animal Species and Evolution." Harvard Univ. Press, Cambridge, Massachusetts.

Meadows, D. H., Meadows, D. L., Randers, J., and Behrens, W. W., III (1972). "The Limits to Growth." Universe Books, New York.

Merrell, D. J. (1951). Interspecific competition between *Drodophila funebris* and *Drosophila melanogaster*. *Am. Nat.* **85**, 159–169.

Mertz, D. B. (1970). Notes on methods used in life-history studies. *In* "Readings in Ecology and Ecological Genetics" (J. H. Connell, D. B. Mertz, and W. W. Murdoch, eds.), pp. 4–17. Harper, New York.

Mertz, D. B. (1971a). Life history phenomena in increasing and decreasing populations. *In* "Statistical Ecology" (G. P. Patil, E. C. Pielou, and W. E. Waters, eds.), Vol. 2, pp. 361–399. Pennsylvania State Univ. Press, University Park.

Mertz, D. B. (1971b). The mathematical demography of the California Condor population. *Am. Nat.* **105**, 437–453.

Mertz, D. B., and Wade, M. J. (1976). The prudent prey and the prudent predator. *Am. Nat.* **110**, 489–496.

Mileikovsky, S. A. (1971). Types of larval development in marine bottom invertebrates, their distribution and ecological significance: A re-evaluation. *Mar. Biol.* **10**, 193–213.

Miller, R. S. (1967). Pattern and process in competition. *Adv. Ecol. Res.* **4**, 1–74.

Milne, A. (1957). The natural control of insect populations. *Can. Entomol.* **89**, 193–213.

Milne, A. (1958a). Theories of natural control of insect populations. *Cold Spring Harbor Symp. Quant. Biol.* **22**, 253–271.

Milne, A. (1958b). Perfect and imperfect density dependence in population dynamics. *Nature (London)* **182**, 1251–1252.

Milne, A. (1959). A controversial equation in population ecology. *Nature (London)* **184**, 1582.

Milne, A. (1961). Definition of competition among animals. *Symp. Soc. Exp. Biol.* **15**, 40–61.

Milne, A. (1962). On a theory of natural control of insect population. *J. Theor. Biol.* **3**, 19–50.

Morris, R. F., Cheshire, W. F., Miller, C. A., and Mott, D. G. (1958). The numerical response of avian and mammalian predators during a gradation of the spruce budworm. *Ecology* **39**, 487–494.

Morton, E. S. (1977). Intratropical migration in the Yellow-green Vireo and Piratic Flycatcher. *Auk* **94**, 97–106.

Mountford, M. D. (1968). The significance of litter-size. *J. Anim. Ecol.* **37**, 363–367.

Murdoch, W. W. (1966). Population stability and life history phenomena. *Am. Nat.* **100**, 5–11.

Murphy, G. I. (1968). Pattern in life history and the environment. *Am. Nat.* **102**, 391–403.

Murphy, R. C. (1936). "Oceanic Birds of South America." Am. Mus. Nat. Hist., New York.

Murray, B. G., Jr. (1967). Dispersal in vertebrates. *Ecology* **48**, 975–978.

Murray, B. G., Jr. (1969). A comparative study of the Le Conte's and Sharp-tailed sparrows. *Auk* **86**, 199–231.

Murray, B. G., Jr. (1971). The ecological consequences of interspecific territorial behavior in birds. *Ecology* **52**, 414–423.

Murray, B. G., Jr. (1975). Distinguishing the woods from the trees, or seeking unity in diversity. *BioScience* **25**, 149.

Murray, B. G., Jr. (1976). A critique of interspecific territoriality and character convergence. *Condor* **78**, 518–525.

Murray, B. G., Jr., and Gill, F. B. (1976). Behavioral interactions of Blue-winged and Golden-winged warblers. *Wilson Bull.* **88**, 231–254.

Nelson, J. B. (1964). Factors influencing clutch-size and chick growth in the North Atlantic Gannet *Sula bassana*. *Ibis* **106**, 63–77.

Nice, M. M. (1957). Nesting success in altricial birds. *Auk* **74**, 305–321.

Nicholson, A. J. (1933). The balance of animal populations. *J. Anim. Ecol.* **2**, 132–178.

Nicholson, A. J. (1954a). An outline of the dynamics of animal populations. *Aust. J. Zool.* **2**, 9–65.

Nicholson, A. J. (1954b). Compensatory reactions of populations to stresses, and their evolutionary significance. *Aust. J. Zool.* **2**, 1–8.

Nicholson, A. J. (1958a). The self-adjustment of populations to change. *Cold Spring Harbor Symp. Quant. Biol.* **22**, 153–173.

Nicholson, A. J. (1958b). Dynamics of insect populations. *Annu. Rev. Entomol.* **3**, 107–136.

Nicholson, A. J. (1959). Density-dependent factors in ecology. *Nature (London)* **183**, 911–912.

Nicholson, A. J., and Bailey, V. A. (1935). The balance of animal populations. Part I. *Proc. Zool. Soc. London* pp. 551–598.

Nolan, V., Jr. (1978). The ecology and behavior of the Prairie Warbler, *Dendroica discolor*. *Ornithol. Monogr.* No. 26.

Norman, F. I., and Gottsch, M. D. (1969). Artificial twinning in the Short-tailed Shearwater *Puffinus tenuirostris*. *Ibis* **111**, 391–393.

Odum, E. P. (1971). "Fundamentals of Ecology," 3rd Ed. Saunders, Philadelphia, Pennsylvania.

Orians, G. H. (1961). The ecology of blackbird (*Agelaius*) social systems. *Ecol. Monogr.* **31**, 285–312.

Orians, G. H., and Collier, G. (1963). Competition and blackbird social systems. *Evolution* **17**, 449–459.

Orians, G. H., and Willson, M. F. (1964). Interspecific territories of birds. *Ecology* **45**, 736–745.

Owen, D. F. (1977). Latitudinal gradients in clutch size: An extension of David Lack's theory.

In "Evolutionary Ecology" (B. Stonehouse and C. Perrins, eds.), pp. 171–179. Univ. Park Press, Baltimore, Maryland.

Paine, R. T. (1966). Food web complexity and species diversity. *Am. Nat.* **100**, 65–75.

Parkes, K. C. (1965). Character displacement in some Philippine cuckoos. *Living Bird* **4**, 89–98.

Paulik, G. J. (1971). Anchovies, birds and fishermen in the Peru Current. *In* "Environment: Resources, Pollution and Society" (W. W. Murdoch, ed.), 1st Ed., pp. 156–185. Sinauer, Stamford, Connecticut.

Peakall, D. B. (1970). The Eastern Bluebird: Its breeding season, clutch size, and nesting success. *Living Bird* **9**, 239–255.

Pearl, R., and Reed, L. J. (1920). On the rate of growth of the population of the United States since 1790 and its mathematical representation. *Proc. Natl. Acad. Sci. U.S.A.* **6**, 275–288.

Pearson, O. P. (1966). The prey of carnivores during one cycle of mouse abundance. *J. Anim. Ecol.* **35**, 217–233.

Perrins, C. (1964). Survival of young swifts in relation to brood-size. *Nature (London)* **201**, 1147–1148.

Perrins, C. M. (1965). Population fluctuations and clutch-size in the Great Tit, *Parus major* L. *J. Anim. Ecol.* **34**, 601–647.

Perrins, C. M., and Jones, P. J. (1974). The inheritance of clutch size in the Great Tit (*Parus major* L.). *Condor* **76**, 225–229.

Perrins, C. M., Harris, M. P., and Britton, C. K. (1973). Survival of Manx Shearwaters *Puffinus puffinus*. *Ibis* **115**, 535–548.

Pianka, E. R. (1970). On *r*- and *K*-selection. *Am. Nat.* **104**, 592–597.

Pianka, E. R. (1978). "Evolutionary Ecology," 2nd Ed. Harper, New York.

Pianka, E. R., and Parker, W. S. (1975). Age-specific reproductive tactics. *Am. Nat.* **109**, 453–464.

Pielou, E. C. (1969). "An Introduction to Mathematical Ecology." Wiley (Interscience), New York.

Pielou, E. C. (1977). "Mathematical Ecology." Wiley (Interscience), New York.

Pitelka, F. A. (1958). Some aspects of population structure in the short-term cycle of the Brown Lemming in northern Alaska. *Cold Spring Harbor Symp. Quant. Biol.* **22**, 237–251.

Pitelka, F. A., Tomich, P. Q., and Treichel, G. W. (1955). Ecological relations of jaegers and owls as lemming predators near Barrow, Alaska. *Ecol. Monogr.* **25**, 85–117.

Plumb, W. J. (1965). Observations on the breeding biology of the Razorbill. *Br. Birds* **58**, 449–456.

Poole, R. W. (1974). "An Introduction to Quantitative Ecology." McGraw-Hill, New York.

Preston, F. W. (1969). Diversity and stability in the biological world. *Brookhaven Symp. Biol.* **22**, 1–12.

Pulliam, H. R. (1974). On the theory of optimal diets. *Am. Nat.* **108**, 59–74.

Pyke, G. H., Pulliam, H. R., and Charnov, E. L. (1977). Optimal foraging: A selective review of theory and tests. *Q. Rev. Biol.* **52**, 137–154.

Reddingius, J. (1971). Gambling for existence. *Acta Biotheor., Suppl. Primum* **20**, 1–208.

Rice, D. W., and Kenyon, K. W. (1962). Breeding cycles and behavior of Laysan and Black-footed albatrosses. *Auk* **79**, 517–567.

Richards, O. W., and Southwood, T. R. E. (1968). The abundance of insects: introduction. *In* "Insect Abundance" (T. R. E. Southwood, ed.), Symposia of the Royal Entomological Society of London, No. 4, pp. 1–7. Blackwell, Oxford.

Ricker, W. E. (1954). Stock and recruitment. *J. Fish. Res. Board Can.* **11**, 559–623.

Ricker, W. E. (1958). Maximum sustained yields from fluctuating environments and mixed stocks. *J. Fish. Res. Board Can.* **15**, 991–1006.

Ricklefs, R. E. (1967). A graphical method of fitting equations to growth curves. *Ecology* **48**, 978–983.

Ricklefs, R. E. (1969). The nesting cycle of songbirds in tropical and temperate regions. *Living Bird* **8**, 165–175.

Ricklefs, R. E. (1970). Clutch size in birds: Outcome of opposing predator and prey adaptations. *Science* **168**, 599–600.

Ricklefs, R. E. (1973). Fecundity, mortality, and avian demography. *In* "Breeding Biology of Birds" (D. S. Farner, ed.), pp. 366–435. Natl. Acad. Sci., Washington, D.C.

Ricklefs, R. E. (1977). On the evolution of reproductive strategies in birds: reproductive efforts. *Am. Nat.* **111**, 453–478.

Ripley, S. D. (1959). Character displacement in Indian nuthatches (*Sitta*). *Postilla* **42**, 1–11.

Root, R. B., and Olson, A. M. (1969). Population increase of the cabbage aphid, *Brevicoryne brassicae*, on different host plants. *Can. Entomol.* **101**, 768–773.

Ross, H. H. (1957). Principles of natural coexistence indicated by leafhopper populations. *Evolution* **11**, 113–129.

Royama, T. (1969). A model for the global variation of clutch size in birds. *Oikos* **20**, 562–567.

Royama, T. (1977). Population persistence and density dependence. *Ecol. Monogr.* **47**, 1–35.

Sadleir, R. M. F. S. (1965). The relationship between agonistic behaviour and population changes in the deermouse, *Peromyscus maniculatus* (Wagner). *J. Anim. Ecol.* **34**, 331–352.

Schaefer, M. B. (1968). Methods of estimating effects of fishing on fish populations. *Trans. Am. Fish. Soc.* **97**, 231–241.

Schaefer, M. B. (1970). Men, birds and anchovies in the Peru Current—Dynamic interactions. *Trans. Am. Fish. Soc.* **99**, 461–467.

Schaffer, W. M. (1974a). Selection for optimal life histories: The effects of age structure. *Ecology* **55**, 291–303.

Schaffer, W. M. (1974b). Optimal reproductive effort in fluctuating environments. *Am. Nat.* **108**, 783–790.

Schaffer, W. M., and Elson, P. F. (1975). The adaptive significance of variations in life history among local populations of Atlantic salmon in North America. *Ecology* **56**, 577–590.

Schifferli, L. (1978). Experimental modification of brood size among House Sparrows *Passer domesticus*. *Ibis* **120**, 365–369.

Schwerdtfeger, F. (1958). Is the density of animal populations regulated by mechanisms or by chance? *Proc. Int. Congr. Entomol., 10th* **4**, 115–122.

Selander, R. K., and Giller, D. R. (1959). Interspecific relations of woodpeckers in Texas. *Wilson Bull.* **71**, 107–124.

Sharpe, F. R., and Lotka, A. J. (1911). A problem in age-distribution. *Philos. Mag.* **21**, 435–438.

Silliman, R. P. (1968). Interaction of food level and exploitation in experimental fish populations. *U.S. Fish Wildl. Serv., Fish. Bull.* **66**, 425–439.

Silliman, R. P., and Gutsell, J. S. (1958). Experimental exploitation of fish populations. *U. S. Fish Wildl. Serv., Fish. Bull.* **58**, 215–252.

Simpson, G. G. (1952). How many species? *Evolution* **6**, 342.

Skellam, J. G. (1951). Random dispersal in theoretical populations. *Biometrika* **38**, 196–218.

Skutch, A. F. (1949). Do tropical birds rear as many birds as they can nourish? *Ibis* **91**, 430–455.

Skutch, A. F. (1961). Helpers among birds. *Condor* **63**, 198–226.

Skutch, A. F. (1966). A breeding bird census and nesting success in Central America. *Ibis* **108**, 1–16.

Skutch, A. F. (1967). Adaptive limitation of the reproductive rate of birds. *Ibis* **109**, 579–599.

Skutch, A. F. (1976). "Parent Birds and Their Young." Univ. of Texas Press, Austin.

Slobodkin, L. B. (1959). Energetics in *Daphnia pulex* populations. *Ecology* **40**, 232–243.

Slobodkin, L. B. (1961). "Growth and Regulation of Animal Populations." Holt, New York.

Slobodkin, L. B. (1968). How to be a predator. *Am. Zool.* **8**, 43–51.

Slobodkin, L. B. (1974). Prudent predation does not require group selection. *Am. Nat.* **108**, 665–678.

Slobodkin, L. B., and Richman, S. (1956). The effect of removal of fixed percentages of the newborn on size and variability in populations of *Daphnia pulicaria* (Forbes). *Limnol. Oceanogr.* **1**, 209–237.

Smith, A. T. (1978). Comparative demography of pikas (*Ochotona*): Effect of spatial and temporal age-specific mortality. *Ecology* **59**, 133–139.

Smith, C. C., and Fretwell, S. D. (1974). The optimal balance between size and number of offspring. *Am. Nat.* **108**, 499–506.

Smith, F. E. (1952). Experimental methods in population dynamics: A critique. *Ecology* **33**, 441–450.

Smith, F. E. (1954). Quantitative aspects of population growth. *In* "Dynamics of Growth Processes" (E. J. Boell, ed.), pp. 277–294. Princeton Univ. Press, Princeton, New Jersey.

Smith, F. E. (1961). Density dependence in the Australian *Thrips*. *Ecology* **42**, 403–407.

Smith, F. E. (1963). Density-dependence. *Ecology* **44**, 220.

Smith, H. S. (1935). The role of biotic factors in the determination of population densities. *J. Econ. Entomol.* **28**, 873–898.

Smith, S. M. (1978). The "underworld" in a territorial sparrow: Adaptive strategy for floaters. *Am. Nat.* **112**, 571–582.

Snow, D. W. (1958). "A Study of Blackbirds." Allen & Unwin, London.

Snow, D. W. (1978). The nest as a factor determining clutch-size in tropical birds. *J. Ornithol.* **119**, 227–230.

Solomon, M. E. (1949). The natural control of animal populations. *J. Anim. Ecol.* **18**, 1–35.

Solomon, M. E. (1957). Dynamics of insect populations. *Annu. Rev. Entomol.* **2**, 121–142.

Solomon, M. E. (1958a). Meaning of density-dependence and related terms in population dynamics. *Nature (London)* **181**, 1778–1780.

Solomon, M. E. (1958b). Perfect and imperfect density dependence in population dynamics. *Nature (London)* **182**, 1252.

Solomon, M. E. (1964). Analysis of processes involved in the natural control of insects. *Adv. Ecol. Res.* **2**, 1–58.

Solomon, M. E. (1969). "Population Dynamics." Arnold, London.

Southern, H. N. (1970). The natural control of a population of Tawny Owls (*Strix aluco*). *J. Zool.* **162**, 197–285.

Spencer, A. W., and Steinhoff, H. W. (1968). An explanation of geographic variation in litter size. *J. Mammal.* **49**, 281–286.

Spinage, C. A. (1972). African ungulate life tables. *Ecology* **53**, 645–652.

Stearns, S. C. (1976). Life-history tactics: A review of the ideas. *Q. Rev. Biol.* **51**, 3–47.

Stearns, S. C. (1977). The evolution of life history traits: A critique of the theory and a review of the data. *Annu. Rev. Ecol. Syst.* **8**, 145–171.

Stefanski, R. A. (1967). Utilization of the breeding territory in the Black-capped Chickadee. *Condor* **69**, 259–267.

Stewart, R. E., and Aldridge, J. W. (1951). Removal and repopulation of breeding birds in a spruce–fir community. *Auk* **68**, 471–482.

Summers-Smith, D. (1956). Mortality of the House Sparrow. *Bird Study* **3**, 265–270.

Talbot, L. M. (1977). Wildlife quotas sometimes ignored the real world. *Smithsonian* **8**(2), 116–118, 120–122, 124.

Tanner, J. T. (1966). Effects of population density on growth rates of animal populations. *Ecology* **47**, 733–745.

Taylor, H. M., Gourley, R. S., Lawrence, C. E., and Kaplan, R. S. (1974). Natural selection of life history attributes: An analytical approach. *Theor. Popul. Biol.* **5**, 104–122.

Thompson, C. F. (1977). Experimental removal and replacement of territorial male Yellow-breasted Chats. *Auk* **94**, 107–113.

Thompson, D. Q. (1955). The role of food and cover in population fluctuations of the brown lemming at Point Barrow, Alaska. *Trans. North Am. Wildl. Conf.* **20**, 166–176.

Thompson, W. R. (1939). Biological control and the theories of the interactions of populations. *Parasitology* **31**, 299–388.

Thompson, W. R. (1956). The fundamental theory of natural and biological control. *Annu. Rev. Entomol.* **1**, 379–402.

Thorson, G. (1950). Reproductive and larval ecology of marine bottom invertebrates. *Biol. Rev. Cambridge Philos. Soc.* **25**, 1–45.

Tilley, S. G. (1973). Life histories and natural selection in populations of the salamander *Desmognathus ochrophaeus*. *Ecology* **54**, 3–17.

Tinbergen, N. (1957). The functions of territory. *Bird Study* **4**, 14–27.

Tinkle, D. W. (1961). Geographic variation in reproduction, size, sex ratio and maturity of *Sternothaerus odoratus* (Testudinata: Chelydridae). *Ecology* **42**, 68–76.

Tinkle, D. W. (1969). The concept of reproductive effort and its relation to the evolution of life histories of lizards. *Am. Nat.* **103**, 501–516.

Tinkle, D. W., Wilbur, H. M., and Tilley, S. G. (1970). Evolutionary strategies in lizard reproduction. *Evolution* **24**, 55–74.

Tompa, F. S. (1964a). Territorial behavior: The main controlling factor of a local Song Sparrow population. *Auk* **79**, 687–697.

Tompa, F. S. (1964b). Factors determining the numbers of Song Sparrows, *Melospiza melodia* (Wilson), on Mandarte Island, B. C., Canada. *Acta Zool. Fenn.* **109**, 3–73.

Tompa, F. S. (1967). Reproductive success in relation to breeding density in Pied Flycatchers, *Ficedula hypoleuca* (Pallas). *Acta Zool. Fenn.* **118**, 1–28.

Tompa, F. S. (1971). Catastrophic mortality and its population consequences. *Auk* **88**, 753–759.

Urdang, L., ed. (1968). "The Random House Dictionary of the English Language. College Edition." Random House, New York.

Utida, S. (1957). Cyclic fluctuations of population density intrinsic to the host–parasite system. *Ecology* **38**, 442–449.

Varley, G. C. (1958). Meaning od density-dependence and related terms in population dynamics. *Nature (London)* **181**, 1780–1781.

Varley, G. C. (1959a). Density-dependent factors in ecology. *Nature (London)* **183**, 911.

Varley, G. C. (1959b). A controversial equation in population ecology. *Nature (London)* **184**, 1583.

Vaurie, C. (1950). Notes on some Asiatic nuthatches and creepers. *Am. Mus. Novit.* No. 1472, 1–39.

Vaurie, C. (1951). Adaptive differences between two sympatric species of nuthatches (*Sitta*). *Proc. Intl. Ornithol. Congr., 10th* pp. 163–166.

Verhulst, P. F. (1838). Notice sur la loi que la population suit dans son accroisissement.

Corresp. Math. Phys. **10**, 113–121. [Partial Engl. transl. *in* "Readings in Ecology" (E. J. Kormondy, ed.), pp. 64–66. Prentice-Hall, Englewood Cliffs, New Jersey, 1965.]

Vermeer, K. (1963). The breeding ecology of the Glaucous-winged Gull (*Larus glaucescens*) on Mandarte Island, B. C. *Occas. Pap., B. C. Prov. Mus. Nat. Hist. Anthropol.* **13**, 1–104.

Volterra, V. (1931). Variation and fluctuations of the number of individuals in animal species living together. *In* "Animal Ecology" (R. N. Chapman), pp. 409–446. McGraw-Hill, New York.

von Haartman, L. (1955). Clutch size in polygamous species. *Acta Int. Ornithol. Congr., 11th, 1954* pp. 450–453.

von Haartman, L. (1967). Clutch-size in the Pied Flycatcher. *Proc. Int. Ornithol. Congr., 14th* pp. 155–164.

von Haartman, L. (1971). Population dynamics. *In* "Avian Biology" (D. S. Farner and J. R. King, eds.), Vol. 1, pp. 391–459. Academic Press, New York.

Wagner, H. O. (1957). Variation in clutch size at different latitudes. *Auk* **74**, 243–250.

Wangersky, P. J., and Cunningham, W. J. (1958). Time lag in population models. *Cold Spring Harbor Symp. Quant. Biol.* **22**, 329–338.

Watson, A. (1967). Population control by territorial behaviour in Red Grouse. *Nature (London)* **215**, 1274–1275.

Watson, A., and Jenkins, D. (1968). Experiments on population control by territorial behaviour in Red Grouse. *J. Anim. Ecol.* **37**, 595–614.

Watson, A., and Moss, R. (1970). Dominance, spacing behaviour and aggression in relation to population limitation in vertebrates. *In* "Animal Populations in Relation to their Food Resources" (A. Watson, ed.)., British Ecological Society Symposium, No. 10, pp. 167–220. Blackwell, Oxford.

Watt, K. E. F. (1955). Studies on population productivity. I. Three approaches to the optimum yield problem in populations of *Tribolium confusum*. *Ecol. Monogr.* **25**, 269–290.

Watt, K. E. F. (1968). "Ecology and Resource Management: A Quantitative Approach." McGraw-Hill, New York.

Watt, K. E. F. (1969). A comparative study on the meaning of stability in five biological systems: Insect and furbearer populations, influenza, Thai hemorrhagic fever, and plague. *Brookhaven Symp. Biol.* **22**, 142–150.

Wharton, C. H. (1966). Reproduction and growth in the cottonmouths, *Agkistrodon piscivorus* Lacépède, of Cedar Keys, Florida. *Copeia* pp. 149–161.

Wiens, J. A. (1966). On group selection and Wynne-Edwards' hypothesis. *Am. Sci.* **54**, 273–287.

Williams, G. C. (1957). Pleiotropy, natural selection, and evolution of senescence. *Evolution* **11**, 398–411.

Williams, G. C. (1966a). "Adaptation and Natural Selection." Princeton Univ. Press, Princeton, New Jersey.

Williams, G. C. (1966b). Natural selection, the costs of reproduction, and a refinement of Lack's principle. *Am. Nat.* **100**, 687–690.

Wilson, E. O. (1975). "Sociobiology." Harvard Univ. Press, Cambridge, Massachusetts.

Wynne-Edwards, V. C. (1955). Low reproductive rates in birds, especially sea-birds. *Acta Int. Ornithol. Congr., 11th, 1954* pp. 540–547.

Wynne-Edwards, V. C. (1962). "Animal Dispersion in Relation to Social Behaviour." Oliver & Boyd, Edinburgh.

Zimmerman, J. L. (1971). The territory and its density dependent effect in *Spiza americana*. *Auk* **88**, 591–612.

SUBJECT INDEX

A

Abundance, 70–73, 187–188
Age of first reproduction
  in anadromous fishes, 106–107
  effect on clutch size, 85
  effect of postponing, 131
  evolution of, 104–105, 120–121
Age distribution, *see also* Stable age distribution
  effect on birth and death rates, 39, 40
  effect on population growth, 35–36, 48
Aggression, *see also* Territorial behavior
  limiting mammal populations, 43
Agkistrodon piscivorus, 93
Agonum, 84
Aloplex lagopus, 56
Alouatta palliata, 164
Anadromous fishes, evolution of life history, 106–107
Anchovy fishery, 148, 149
Aneides lugubris, 93
Aphid, 29, 125–126
Area effect
  on community composition, 7
  on competition, 178–179
  on distribution and abundance, 7

on population dynamics, 7–9
on territoriality, 44–45
Assumptions
  for analysis of population dynamics, 7, 21, 28, 36
  for evolution of life history patterns, 82–84
  for food-limited population model, 46
  for space-limited population model, 38
Atlantic Puffin (*Fratercula arctica*), 115

B

b, birth rate, 21, 22
b_r, replacement rate, 24, 26, 185
  as index of survivorship, 37
  of intermittently breeding populations, 26
Balanus balanoides, 172, 174
Balanus glandula, 179
Batrachoseps attenuatus, 93
Black-winged Red Bishop (*Euplectes hordeacea*), 172
Blue Tit (*Parus caerulea*), 100
Breeding, *see* Intermittent breeding

Breeding season
  effect of length on population growth, 2,
    26, 59–61, 186
  timing of, 96–97
Brevicoryne brassicae, 185
Brown Booby (*Sula leucogaster*), 113

C

Cactoblastis cactorum, 8–9, 14, 163, 187
Calandra oryzae, 29, 126
Callosobruchus chinensis, 178
Carnivore, 45, 50, 54
Carrying capacity, 1, 2, 15
Centropus, 177
Character displacement, 167, 175–177
Choristoneura fumiferana, 55
Chrysemys picta, 91
Chthamalus stellatus, 172, 174
Clutch size, *see also* Fecundity; Reproduc-
    tion
  annual variation, 95–97
  defined, 86
  density-dependence of, 11–12
  effect of age of first breeding, 85, 101
  effect of annual survivorship, 83–85, 89
  effect of climate, 92–93
  effect of development time, 87–89
  effect of length of breeding season,
    87–89
  effect of length of larval life, 92
  effect of nest size, 120
  effect of parental care, 103–104, 118
  genetic variation, 110–112
  geographic variation, 87–95
  heritability, 110
  latitudinal variation, 89–92
  as outcome of predator–prey adapta-
    tions, 120
  and population dynamics, 121–122
  and primary productivity, 101–103, 118
  and reproductive effort, 79
  seasonal variation, 96–97, 98
  variation with body size, 97, 100–101, 102
  variation with elevation, 93–94
  variation with food supply, 96
Coefficient of annual increase (r'_a), 24, 27,
    61, 73
Coexistence, 165, 166, 169, 179
  of *Drosophila*, 178, 179, 180

of parasitic wasps, 178
of prey populations, 179
Colinus virginianus, 6
Common Swift (*Apus apus*), 108–109, 115
Compensation, 147
Competition, *see also* Interspecific compe-
    tition; Intraspecific competition;
    and Territorial behavior
  defined, 164
  equations, 2, 165
  meaning of Nicholson, 64
  without shortages, 181–182
Competitive exclusion
  of barnacles, 172, 174
  of paramecia, 171–172, 173
  of planarians, 171
Competitive exclusion principle
  misapplication, 168–171
  testability, 177
  as third law of population dynamics, 164
Crypticity, effect of selection for, 130

D

Daphnia obtusa, 84, 151
Daphnia pulex, 25, 84, 185
Daphnia pulicaria, 151
Death rate (d), 21, 185
Decomposer, 45, 48
Density-dependent factors
  effective, 11
  imperfect, 2
  inadequate, 1
  ineffective, 11, 12
  inverse, 11
  perfect, 2
  role in population growth, 1–3, 66–68,
    184
Density-dependent regulation
  criticism of, 68–70
  defense on logical grounds, 3
  lack of evidence, 3, 9
Density-governing reaction, 74
Density-independent factors, 11, 12
Dependence
  causal, 13
  statistical, 13
Desmognathus ochrophaeus, 93
Development time, *see* Clutch size
Disease, 1, 73

Dispersal, 37, 42
Dispersion, 8
Distribution, 70–73, 187–188
Drosophila funebris, 178
Drosophila melanogaster, 178
Drosophila pseudoobscura, 179–180
Drosophila serrata, 179–180
Drosophila subobscura, 84
Dusky Flycatcher (*Empidonax wrightii*),
172

E

Eastern Bluebird (*Sialis sialia*), 105
Eastern Meadowlark (*Sturnella magna*),
172
Economics of exploitation, 161
Emigration, 5, 71, 130
Environment
defined, 15
effect on distribution and abundance, 72
effect on survivorship and reproduction,
72, 187
Eotetranychus sexmaculatus, 9, 10, 187
Equilibrium system, 36, 184
European Blackbird (*Turdus merula*), 84,
109
European Robin (*Erithacus rubecula*), 92
Exponential equation, 19
meaning of r in, 124
Exponential growth, 1, 69
Exponential growth rate, 47
Exponential model, 48
Extinction, 13–14, 24, 184

F

Famine, 45
Fecundity, *see also* Clutch size; Reproduction
in animals with determinate growth,
85–86
in animals with indeterminate growth,
86, 101
defined, 4
effects on MSY, 141
relationship with survivorship, 86–87
variation, 185
First law of population dynamics, 30, 59,
70, 125

Fisheries management
dynamic pool models, 134
logistic-type models, 134
Floaters, 39, 42, 43
Fluctuations, 4–5, 6, 15–16
considered stable, 13–14
magnitude, 53–54, 186
of no theoretical interest, 5, 6
Food
as independent variable, 68
as limiting factor, 45–58
as regulating factor, 58–59
Food-limited population, 71, 185, 186
clutch size, 103
effect of natural selection on, 129–131
effect of predation on, 52–54, 135–141
Food-limited population model, 46–47
compared with logistic model, 47–48,
49–50
Functional response, 57

G

Galapagos finches, 177
Galileo, 177
Gambling for existence, 3
Generation time (T), 21, 22
relationship with body size, 97, 99, 124
relationship with intrinsic rate of increase, 99–100, 124
Geomys, 91
Glaucous Gull (*Larus hyperboreus*), 56
Glaucous-winged Gull (*Larus glaucescens*), 114
Golden-fronted Woodpecker (*Centurus aurifrons*), 172
Gray Flycatcher (*Empidonax oberholseri*),
172
Great Tit (*Parus major*)
clutch size, 11, 12, 66, 94–95
foraging, 57
nestling weight, 109
population fluctuations, 16
Group selection, 119, 134, 160, 184
Growth rate, *see also* r
individual, 144–145
population, 80
Gymnopais, 91

H

Habitat
    ephemeral, 125
    heterogeneous, 165
    quality, 2, 70–71, 126–127
Hardy–Weinberg law, 30
Herbivore, 45, 50, 54
Herring Gull (*Larus argentatus*), 112, 114
Hole-nesting
    effect on age of first reproduction, 104–
        105
    limiting population size, 43
House Finch (*Carpodacus mexicanus*),
    115, 119
House Martin (*Delichon urbica*), 115
House Sparrow (*Passer domesticus*), 84,
    116–117

I

Icterine Warbler (*Hippolais icterina*), 172
Immigration
    counted as recruitment, 71
    effect on survivorship schedule, 5
Intermittent breeding, 63
    effect on annual rate of increase, 24–28
    with respect to r-selection, 126
Interspecific competition, 73, *see also*
        Competitive exclusion
    consequences of, 164–170
Interspecific territoriality, *see* Territorial
        behavior
Intraspecific competition, 2, 73, *see also*
        Territorial behavior
    for food, 45–58, 68, 71, 185
    as mechanism reducing growth rate, 74
    for resources, 1
    for space, 38–45
Iteroparity
    in Atlantic salmon, 107
    conditions for, 106, 121
    origin of, 80

K

K-selection, 2, 76, 122–128
    correlates of, 88
K strategist, 124

L

l_x, *see also* Survivorship
    defined, 20, 76–77
    distinguished from λ_x, 77
    as net effect of deaths and migration, 37
    sampling errors, 28
λ_x, defined, 77
Lacerta, 93
Lack's hypothesis, 108–118
    criticism of, 108, 118
Laysan Albatross (*Diomedea immutabilis*),
    113
Lebistes reticulatus, 48, 52, 151–156, 157,
    158, 159
Lemmus trimucronatus, 56
Leslie matrix, 2, 34
Lesser Black-backed Gull (*Larus fuscus*),
    112, 114
Life expectancy, 30, 95–96, 101
Life history patterns, evolution of, 81–121
Limitation
    distinguished from regulation, 2, 14–15
    theories, 38
Limiting factor
    food, 45–58
    length of breeding season, 60
    length of nonbreeding season, 61
    predation, 51–58
    setting carrying capacity, 14
    space, 38–45
    territorial behavior, 38–45
    time, 59–66
    weather, 61–66
Litter size, 86, *see also* Clutch size
Lizards, 87, 91, 101
Logistic equation, 2, 19–20
    advantages of, 20
    assumptions of, 20, 24
    criticisms of, 2, 19–20
    limitations of, 24
    meaning of r in, 124
    modeling fluctuations, 13
Logistic growth model, compared with
        food-limited model, 47–48, 68
Lotka's equations, 20–24, 125
Lotka–Volterra competition equations,
    165–167
    criticisms of, 165–166

Lower critical density (LCD), 46
Lucilia cuprina, 13, 14, 84, 151

M

m_x, *see also* Fecundity; Recruitment
    defined, 20, 76–77, 86
    distinguished from μ_x, 77
    sampling errors, 29
μ_x, defined, 77, 86
Malthusean parameter (m), 21, 77
Manx Shearwater (*Puffinus puffinus*), 113
Marmota flaviventris, 119
Maximum sustainable yield (MSY)
    of biomass, 143–146
    epitaph of, 133
    factors affecting, 141, 147–151
    of food-limited populations, 135–141
    of numbers, 141–143
    of prudent predator, 160–161
Meadow Pipit (*Anthus pratensis*), 93
Melodious Warbler (*Hippolais polyglotta*),
    172
Microtus agrestis, 26, 29, 126
Mitella polymerus, 179
Models
    ambiguity of, 81
    comparison of, 47
    confrontation of mathematicians and
        biologists over, 18
    evaluation of, 3, 13
Mortality, sources of data, 28
Mustela rixosa, 56
Mytilus californianus, 179

N

Natural selection, 76, 128, 179, 182, 183,
    184
    consequences for population dynamics,
        81, 128–132
    of life history patterns, 81–107
    of traits affecting survivorship and re-
        production, 79
Neocatoloccus mamezophagus, 178
Net reproductive rate (R_0), 21
    as function of body size, 100, 124
    as function of intrinsic rate of increase,
        100, 124

Newton's first law of motion, 30
Nonbreeding season, effects of, 60–61
North Atlantic Gannet (*Sula bassana*),
    112, 113, 118
Numerical response, 55, 57

O

Ocenebra, 84
Ochotona princeps, 94
Onchorhynchus, 107
Oncopeltus fasciatus, 29, 84, 185
Optimal foraging theory, 56–57
Optimal sustainable population, 133
Opuntia, 8–9, 14, 163, 187
Ostrich (*Struthio camelus*), 100
Overcrowding, 48
Overharvesting, 133, 146, 150, 160
Overpopulation, 37
Owls, clutch size, 91

P

Papio anubis, 84
Paramecium aurelia, 171–172, 173
Paramecium caudatum, 171–172, 173
Parasitic Jaeger (*Stercorarius parasiticus*),
    56
Pediculus humanus, 29
Periplaneta americana, 84
Peromyscus maniculatus, 94
Perturbations, 36
Pest control, 161–163
Physa gyrina, 126
Pied Flycatcher (*Muscicapa hypoleuca*),
    109
Pimephales promelus, 84
Pisaster ochraceus, 179
Planaria gonocephala, 171
Planaria montenegrina, 171
Pomarine Jaeger (*Stercorarius
    pomarinus*), 56, 96
Population
    defined, 15
    density, numbers, and size as equivalent
        terms, 7
    food-limited, 45–55, 68, 71, 162, 185, 186
    herbivore-limited, 46
    limitation, 38

pathogen-limited, 162
predator-limited, 46, 51–58, 68, 162, 185, 186
regulated, 1, 2, 3, 58–59, 74–75
residual, 143, 144
space-limited, 71, 185, 186
time-limited, 59–62, 67, 163, 185, 186
Population growth, 4–5
Population parameters, geographic variation, 4
Population statistics, calculation, 22–23
Prairie Warbler (*Dendroica discolor*), 101
Predation
effect on competing species, 148, 179
effect on population parameters, 1, 135, 136, 137, 138, 139, 140, 144
nonlimiting, 54, 58
prudent, 134, 160–161
Predator-prey interactions, 50–58, 134–141, 147–148, 160–161
functional response, 57
numerical response, 55, 57
Pristiphora erichsonii, 54–56
Producer, 45, 48
Projection
conditions for, 6, 19
matrix, 34
Pseudemys scripta, 91

R

r, rate of increase, 19, 21, 123–124
distinguished from ρ, 77
of intermittently breeding population, 25, 27
r-selection, 2, 76, 122–128
correlates of, 88
r strategist, 124
ρ, defined, 77
Raven (*Corvus corax*), 91
Razorbill (*Alca torda*), 114, 115
Recruitment, 30, 32, 42
effect on MSY, 150
effect of one time change in, 35–36, 186
in harvested populations, 144, 146
limits population growth, 31
Red-bellied Woodpecker (*Centurus carolinensis*), 172
Red Crossbill (*Loxia curvirostra*), 91
Redshank (*Tringa totanus*), 57

Red-winged Blackbird (*Agelaius phoeniceus*), 104, 174–175
Regulating factor, 75
Regulating mechanism
food as, 58–59
predation as, 57–58
Regulation
distinguished from limitation, 14–15
meaning of, 9
theories, 38
Reproduction
adjusted, 118–119
costs of, 84
maximal, 107–118
optimal, 119
Reproductive effort, 79
Reproductive value, 77–79, 160
Rock nuthatches (*Sitta neumayer* and *S. tephronota*), 176–177

S

Salmo salar, 107
Scavenger, 45, 48
Science, goal of, 17–18
Scientific method, 79–80
Second law of population dynamics, 31, 33, 34, 39
Second law of thermodynamics, 45
Self-regulation, 74
Semelparity, evolution of, 105–107, 120–121
Senescence, evolution of, 83
Shaskyus, 84
Short-eared Owl (*Asio flammeus*), 56, 96
Short-tailed Shearwater (*Puffinus tenuirostris*), 113
Snowy Owl (*Nyctea scandiaca*), 56, 96
Social species, effects of predation on, 148, 150
Song Sparrow (*Melospiza melodia*), 42–43, 93
Song Tanager (*Ramphocelus passerinii*), 119
Song Thrush (*Turdus ericetorum*), 109
South Atlantic Gannet (*Sula capensis*), 112, 113
Space-limited populations, 38–45, 67, 186, *see also* Territorial behavior
effect of natural selection on, 130–131

Species-area curve, 7
Spermophilus, 91
Spotted Flycatcher (*Muscicapa striata*), 57
Stability
  defined, 13–14
  habitat, 123, 126–127
  neutral, 13
Stable age distribution, 21, 22, 23, 24,
    33–34, 48, 185, *see also* Age distribu-
    tion; Second law of population
    dynamics
Starling (*Sturnus vulgaris*), 109, 116–117
Stationary-state, 30
Steady-state, 4–5, 22, 30, 186
  age distribution, 23, 24
  conditions for, 30–31, 59
  distinguished from "steady-state," 70
  size affected by perturbation, 36
Sternothaerus odoratus, 91
Struggling for existence, 3, 183
Survivorship, *see also* Clutch size; l_x
  effect on MSY, 141, 150
  in natural situations, 103, 185
  relationship with fecundity, 86–87
  variation, 185
Survivorship schedule
  horizontal, 6
  of marked cohort, 6
  as net effect of deaths and migration, 37
  vertical, 6
Swallow-tailed Gull (*Creagrus furcatus*),
    112, 114

T

Tawny Owl (*Strix aluco*), 96
Territorial behavior, *see also* Space-
    limited populations
  effects on birth and death rates, 39, 40
  effects on population growth, 38–39
  and group selection theory, 44
  interspecific, 172, 174–175
  limiting annual recruitment, 38
  as limiting factor, 38–45
  limiting population size, 38–39, 41, 54
  prevents breeding, 42
  with respect to population density, 44–45
Territory size
  effect on birth and death rates, 67

effect of selection on, 130–131
  limiting population size, 67
  role according to Lack, 40
  variation, 41, 42–43, 186
Theories, evaluation of, 73–75
Thick-billed Nutcracker (*Nucifraga
    caryocatactes*), 96
Third law of population dynamics, 164
Thomomys, 91
Thrips imaginis, 62–66
Time, as a limiting factor, 54, 59–66, 163,
    186
Tribolium castaneum, 29, 126
Tribolium confusum, 48, 51, 151
Tricolored Blackbird (*Agelaius tricolor*),
    174
Twinning experiments, 112–118
  criticisms of, 117–118
Typhlodromus occidentalis, 9, 10, 187

U

Undercrowding, 46, 48
Underharvesting, 150
Upper critical density (UCD)
  effect of natural selection on, 129
  in food-limited populations, 46–48, 68,
    71, 73, 135
  in pest populations, 163
  relationship with MSY, 140–147, 151–
    153, 156, 161
  variation with abundance of resources,
    46–47, 103

W

Weather
  as density-dependent factor, 11, 63
  as density-independent factor, 2, 12
  determining length of breeding season,
    89
  effect on annual rate of increase, 61, 67,
    73
  as limiting factor, 60, 61–66
  multiple effects on population numbers,
    60, 62
Western Meadowlark (*Sturnella neglecta*),
    172

White-bearded Manakin (*Manacus man-acus*), 115
White Booby (*Sula dactylatra*), 113

Y

Yellow-eyed Junco (*Junco phaeochroa*), 115, 119
Yellow-headed Blackbird (*Xanthocephalus xanthocephalus*), 174–175

Yield, *see also* Maximum sustainable yield (MSY)
  of fluctuating populations, 150–151
  of harvested populations, 141–146, 151–161

Z

Zanzibar Red Bishop (*Euplectes nigroventris*), 172